Environmental Impact of
MINING AND MINERAL PROCESSING

Environmental Impact of
MINING AND MINERAL PROCESSING
Management, Monitoring, and Auditing Strategies

RAVI K. JAIN, Ph.D., P.E.

Dean Emeritus, School of Engineering and Computer Science, University of the Pacific, Stockton, CA

ZENGDI "CINDY" CUI, Ph.D.

Professor, School of Resources and Earth Science, China University of Mining and Technology, P.R. China

Researcher and Marketing Manager, School of Engineering and Computer Science, University of the Pacific, Stockton, CA

JEREMY K. DOMEN, M.S.

Research Associate, School of Engineering and Computer Science, University of the Pacific, Stockton, CA

ELSEVIER

Amsterdam • Boston • Heidelberg • London
New York • Oxford • Paris • San Diego
San Francisco • Singapore • Sydney • Tokyo

Butterworth-Heinemann is an imprint of Elsevier

Butterworth-Heinemann is an imprint of Elsevier
The Boulevard, Langford Lane, Kidlington, Oxford OX5 1GB, UK
225 Wyman Street, Waltham, MA 02451, USA

Notices
Knowledge and best practice in this field are constantly changing. As new research and
experience broaden our understanding, changes in research methods, professional
practices, or medical treatment may become necessary.

Practitioners and researchers must always rely on their own experience and knowledge in
evaluating and using any information, methods, compounds, or experiments described
herein. In using such information or methods they should be mindful of their own safety
and the safety of others, including parties for whom they have a professional
responsibility.

To the fullest extent of the law, neither the Publisher nor the authors, contributors, or
editors, assume any liability for any injury and/or damage to persons or property as a
matter of products liability, negligence or otherwise, or from any use or operation of any
methods, products, instructions, or ideas contained in the material herein.

ISBN: 978-0-12-804040-9

British Library Cataloguing in Publication Data
A catalogue record for this book is available from the British Library

Library of Congress Cataloging-in-Publication Data
A catalog record for this book is available from the Library of Congress

For Information on all Butterworth-Heinemann publications
visit our website at http://store.elsevier.com/

Working together
to grow libraries in
developing countries

www.elsevier.com • www.bookaid.org

CONTENTS

About the Authors   vii
Preface   ix
Acknowledgements   xi
Acronym List   xiii

1. **Introduction**   **1**
   1.1 Sustainable Development in Mining and Mineral Processing   1
   1.2 Challenges Related to Sustainable Development   3
   1.3 Mining Industry Trends   5
   1.4 Mining Contribution Index   8
   1.5 Mining Processes   12
   1.6 Objectives   13
   References   15

2. **A Systematic Procedure for Environmental Impact Analysis**
   of Mining and Mineral Processing Projects   **17**
   2.1 Introduction   17
   2.2 Systematic Procedure for Preparing Environmental Assessment
       Documentation   18
   References   33

3. **Environmental Management System Implementation in the**
   Mining Industry   **35**
   3.1 Introduction   35
   3.2 What is an EMS?   36
   3.3 Benefits of Implementing an EMS and Best Practices (BPs)   39
   3.4 Costs of Implementing an EMS and BPs   41
   3.5 Government Involvement in EMS   42
   3.6 Implementation of EMS   45
   References   50

4. **Environmental Impacts of Mining**   **53**
   4.1 Introduction   53
   4.2 Air Quality Impacts from Mining   53
   4.3 Water Quality and Quantity Impacts from Mining   91
   4.4 Acid Mine Drainage   111
   4.5 Land Impacts from Mining   117

4.6 Ecological Impacts from Mining   130
4.7 Economic Impacts from Mining   136
References   149

5. **Environmental Monitoring**   **159**
   5.1 Introduction   159
   5.2 Design of Monitoring Plans   161
   5.3 Data Management   186
   5.4 Monitoring Technologies   189
   5.5 Emerging Monitoring Technologies   196
   References   197

6. **Environmental Auditing**   **201**
   6.1 Introduction   201
   6.2 Types of Environmental Audits   201
   6.3 Performing an Environmental Audit   202
   6.4 Standards for Environmental Auditing   208
   6.5 Auditing System Checklists   209
   References   227

7. **Mitigation Measures and Control Technology for Environmental**
   and Human Impacts   **229**
   7.1 Introduction   229
   7.2 Air Pollution Mitigation and Control   229
   7.3 Mitigation of Water Quantity and Quality Impacts   240
   7.4 Mitigation of Land Impacts   247
   7.5 Mitigation of Ecological Impacts   251
   7.6 Fire and Explosion Control   256
   7.7 Human Health and Safety   259
   References   265

Appendix A: Emission Factors for Air Pollutants Related to Mining and Mineral Processing   *271*
Index   *299*

ABOUT THE AUTHORS

Ravi K. Jain, Ph.D., P.E., DEE, Dean Emeritus, was Professor and Dean School of Engineering and Computer Science, University of the Pacific, Stockton, California from 2000 to 2013. Prior to this appointment, he has held research, faculty, and administrative positions at the University of Illinois (Urbana-Champaign), Massachusetts Institute of Technology (MIT), and the University of Cincinnati. He has appointment as distinguished professor at several universities internationally.

Dr. Jain has served as Chair, Environmental Engineering Research Council, American Society of Civil Engineers (ASCE) and is an elected Diplomate of the American Academy of Environmental Engineers (DEE), fellow ASCE and fellow American Association for the Advancement of Science (AAAS). Dr. Jain was the founding Director of the Army Environmental Policy Institute; he has directed major research programs for the U.S. Army and has worked in industry and for the California State Department of Water Resources. He has been a Littauer Fellow at Harvard University and elected a Fellow of Churchill College, Cambridge University. He has published 19 books and more than 180 journal papers, technical reports and book chapters.

He received his B.S. and M.S. degrees in Civil Engineering from California State University, and a Ph.D. in Civil Engineering from Texas Tech University. He studied Public Administration and Public Policy at Harvard University earning an M.P.A. degree. He did additional graduate studies at Massachusetts Institute of Technology (MIT).

Zengdi "Cindy" Cui, Ph.D., is a Professor at the School of Resources and Earth Science, China University of Mining and Technology, China, and also a Researcher and Marketing Manager at the School of Engineering and Computer Science, University of the Pacific, Stockton, California. She has held research, technology transfer, teaching, and administrative positions at these universities. She has appointments as distinguished professor at several universities internationally.

Dr. Cui is a member of the China Coal Association, the International Rock Mechanics Association, and the Urban Geoenvironment & Sustainable Development Center of Ministry of Education, China. She played a key role in the establishment of the Pacific Resources Research Center at the University of the Pacific. She has published several peer reviewed journal articles, book chapters, technical reports and business plans.

Dr. Cui received a B.S. degree from the School of Resources and Earth Science, China University of Mining and Technology, a B.S. degree from the School of Computer Science, Windsor University, a M.S. degree from the School of Computer Science, Wayne State University, a Ph.D. from the School of Resources and Environment, Shandong University of Science and Technology. She also received her MBA from the School of Business, Shandong University of Science and Technology.

Jeremy K. Domen, M.S., has a broad range of experience in research related to water quality, hydraulic fracturing, environmental impact analysis, and engineering innovation. He has published multiple peer-reviewed papers and technical reports and has presented research work at various technical conferences. Mr. Domen has held research appointments at Lawrence Berkeley National Laboratory in Berkeley, California as well as at the School of Engineering and Computer Science at the University of the Pacific, Stockton, California. He participated in a highly selective international engineering internship program, learned the Japanese language and culture and worked in Japan for six months. Mr. Domen received his B.S. degree in Bioengineering and M.S. degree in Engineering Science from the University of the Pacific, School of Engineering and Computer Science.

PREFACE

The National Academy of Engineering (NAE) (2010) described the importance of mining and mineral extraction by stating that the history of human civilization is often characterized by terms such as: Stone Age, Bronze Age, Industrial Revolution, and Information Age. As one can see, a common thread among all these epochs is the extraction of, processing, and utilization of materials from the Earth (NAE, 2010). In fact, almost every product and service in the modern world relies on the raw materials generated by mining and mineral processing. Clearly, mining and mineral extraction have significantly contributed to the advancement of human civilization and national economies. These activities also have the potential for serious environmental impacts. Through the development of best management practices with sustainable development in mind, environmental threats from mining and mineral processing can be minimized as described in this book.

The World Economic Forum (2014) has identified several driving forces toward sustainability and, in response to these drivers, the Forum has also identified major aspects of sustainable development that need to be focused on in relation to mining and mineral processing. To move toward sustainable development, NAE (2010) also noted scientific and technical challenges that need to be overcome. These challenges and issues are described in detail in this book.

Mining and mineral processing are important to the economy of many nations. A comprehensive, interesting, and useful analysis provides information about the importance of these activities for major mining and mineral processing countries. For major mining countries, this analysis provides information such as total mineral export contribution, total production value, and production value as a percentage of the nation's GDP. For comparative analysis purposes, the mining contribution index (MCI) is provided that, in a way, shows the relative importance of these activities to a given country.

Ravi K. Jain
Stockton, California, U.S.A.

Zengdi (Cindy) Cui
Xuzhou, China

Jeremy K. Domen
Stockton, California, U.S.A.

REFERENCES

National Academy of Engineering (NAE), 2010. Grand Challenges for Earth Resources Engineering. National Academy of Engineering. Retrieved from: https://www.nae.edu/File.aspx?id=106323.

World Economic Forum Mining & Metals Industry Partnership, Accenture, 2014. Scoping Paper: Mining and Metals in a Sustainable World. World Economic Forum. Retrieved from: http://www3.weforum.org/docs/WEF_MM_MiningMetalSustainableWorld_ScopingPaper_2014.pdf.

ACKNOWLEDGEMENTS

The authors want to express their deep gratitude for the support in preparing the book manuscript that was provided by **Mr. Xuejun Zhang, Chairman, JingAn Century Investment, Inc.** Mr. Zhang is a visionary industry leader who is keenly interested in environmental, human health, and effective management of mining activities. Support for related projects provided by Shanghai EcoGeological Engineering Inc. is also gratefully acknowledged.

Professor William T. Stringfellow (University of the Pacific and Lawrence Berkeley National Laboratory) reviewed the book manuscript and made numerous comments to further improve the book content and quality. Natalie Muradian, research associate, contributed to the early drafts of the book manuscript. We are grateful to Professor Stringfellow and Ms. Muradian for their contributions.

Many individuals at Elsevier were most helpful with finalizing the manuscript and producing the text. Considerable assistance and support provided by Kenneth P. McCombs, Senior Acquisitions Editor, made the crucial difference in effectively completing this text on a timely basis.

ACRONYM LIST

AMD	Acid metalliferous drainage/Acid mine drainage
ASGM	Artisanal and small-scale gold mining
BACI	Before-After-Control-Impact
BOD	Biological oxygene demand
BP	Best practices
BTEX	Benzene, toluene, ethylbenzene, xylene
CEMS	Continuous emissions monitoring system
COD	Chemical oxygen demand
CT	Communication and tracking
CMM	Coal mine methane
CWA	Clean Water Act
DInSAR	Differential interferometric synthetic aperture radar
EA	Environmental assessment
EIA	Environmental impact assessment
EIS	Environmental impact statement
EMS	Environmental management systems
EPA	Environmental Protection Agency
FONSI	Finding of no significant impact
GDP	Gross domestic product
GIS	Geographic information system
ICMM	International Council on Mining & Metals
IPCC	Intergovernmental Panel on Climate Change
ISO	International Organization for Standardization
MCI	Mining Contribution Index
MINER Act	Mine Improvement and New Emergency Response Act
MSHA	Mine Safety and Health Administration
NAE	National Academy of Engineering
NGO	Non-governmental organization
NIOSH	National Institute for Occupational Safety and Health
NPDES	National Pollution Discharge Elimination System
OAIMA	Ohio Aggregates and Industrial Minerals Association
OEM	Original equipment manufacturer
ORP	Oxidation reduction potential
PAH	Polycyclic aromatic hydrocarbons
PM$_{10}$	Particulate matter with diameter <10 microns
PM$_{2.5}$	Particulate matter with diameter <2.5 microns

PM	Particulate matter
PPE	Personal protective equipment
QA/QC	Quality assurance/quality control
R&D	Research and development
RFID	Radio-frequency identification
TDS	Total dissolved solids
TSS	Total suspended solids
TSF	Tailings storage facility
UHF	Ultra high frequency
VHF	Very high frequency
VOC	Volatile organic compound
WEPP	Water Erosion Prediction Project
WEPS	Wind Erosion Prediction System
WEQ	Wind Erosion Equation

CHAPTER 1

Introduction

1.1 SUSTAINABLE DEVELOPMENT IN MINING AND MINERAL PROCESSING

The importance of mining and mineral extraction is best characterized by the National Academy of Engineering (NAE) (2010), where it was stated that the history of human civilization is often characterized by periods such as the Stone Age, Bronze Age, Industrial Revolution, and Information Age. As can be seen, a common thread among all these epochs is the extraction of, processing, and utilization of materials from the earth (NAE, 2010). For example, materials such as iron and coal fueled the industrial revolution, hydrocarbons and fertilizers fueled recent economic and population growth, and rare earth elements have been critical to the development of modern electronics (NAE, 2010). While mining and mineral extraction have significantly contributed to the advancement of human civilization and national economies, they also have the potential for serious environmental degradation. Through the development of best management practices with sustainable development in mind, environmental threats from mining and mineral processing can be minimized.

The United Nations World Commission on Environment and Development (1987) defines sustainable development as "meet[ing] the needs of the present generation without compromising the ability of the future generation to meet their needs." Sustainable development can also be viewed as a process that "involves the economic, social, cultural and environmental dimensions of human existence" (United Nations, 2002). Another related concept developed by John Elkington in 1994 suggests an appropriate balance is needed between economic prosperity, environmental quality, and social justice.

As the world moves toward more sustainable practices, the mining and mineral processing industry will be profoundly impacted. Mining provides the raw materials found in virtually every product and service throughout the world. However, the long project life cycles for mining operations means companies are accustomed to long-term plans and operations are generally static (World Economic Forum Mining & Metals Industry

Environmental Impact of Mining and Mineral Processing
ISBN 978-0-12-804040-9
http://dx.doi.org/10.1016/B978-0-12-804040-9.00001-2

Partnership and Accenture, 2014). To assist the mining industry in the coming transition to a more sustainable world, the World Economic Forum (2014) has identified several driving forces toward sustainability that will impact the mining industry:

1. Increased demand for fairness: Demand for more equal distribution of benefits and risks between the community, government, stakeholders, and industry
2. Increased "democratization": Demand for greater transparency, regulations, and standards to create a level playing field
3. Increased environmental concerns: Renewed focus on the sustainable management of the environment, climate, water resources, and biodiversity
4. Generational changes: New values will begin to emerge in leadership, government, and society
5. Rapid development of technology: New technologies will transform current operations and processes
6. Shifts in global production: Mining and production will increase in remote, undeveloped, and previously inaccessible areas
7. Increased concerns about artisanal and small-scale mining: Holding small-scale operations accountable to the same standards as large operations

In response to these drivers, the International Council on Mining & Metals (2013), the World Economic Forum (2014), and the World Coal Association (2014) have identified major aspects of sustainable development that need to be focused on in relation to mining and mineral processing:

- Investing in research and development of new technologies
- Encourage re-use, recycling, and responsible disposal of waste products
- Adopting management strategies based on collected data and sound science
- Continual reduction of environmental impacts and protection of biodiversity
- Continual reduction of emissions
- Protecting water resources
- Commitments to health and safety, human rights, cultural preservation, fair employment, and employee training
- Contributing to the social, economic, and institutional well-being of local communities
- Engaging communities, stakeholders, and governments
- Fostering trustworthiness, transparency, and ethical business practices

- Incorporate sustainable development ideas in the corporate decision-making process
- Maintaining good governance

Despite the identification of areas that require improvement by the mining and mineral processing industry, there remain significant challenges related to sustainable development.

1.2 CHALLENGES RELATED TO SUSTAINABLE DEVELOPMENT

To move toward sustainable development, NAE (2010) noted the following scientific and technical challenges that need to be overcome:

1. Making the Earth "transparent"
   a. Because the Earth is solid, it is difficult to understand processes that occur underground. Improved tools are needed to discover, delineate, and identify subsurface resources, to be able to identify the flow of fluids and contaminants in the subsurface, and to detect and monitor fractures in rock structures. Current technologies that contribute to the improved "transparency" include geophysical examinations, a wide range of survey techniques that allow everything from large-scale investigations to microscopic investigations. These techniques include large scale (e.g., airborne, electromagnetic, and magnetic imaging and synthetic aperture radar measurements), medium scale (e.g., seismic waves and pumping and tracer tests), and small scale techniques such as well logging. Imaging technology such as 3D X-ray tomography scanning, 3D scanning electron microscope imaging, and acoustic and confocal microscopy are useful when researching fluid flows in the subsurface on a small scale.
   b. These technologies will enable a more accurate and precise estimate of the Earth's resources. They will also increase safety and decrease risk by increasing the knowledge of fracture locations, stability of structures, and potential contamination pathways.
2. Understanding, engineering, and controling subsurface processes
   a. Underground mechanical, biological, chemical, thermal, and hydrological processes are highly complex and are often coupled together. Understanding how these processes interact is further complicated by the effects of differing time scales and physical dimensions. As numerical simulation models depend on the accuracy of the equations used to simulate these interactions, model results can only be

as good as the current understanding of interactions. A better understanding of these interactions will benefit the exploration of minerals, surface and underground mining, in-situ mining operations, predicting transport and fate of contaminants in groundwater, and earthquake mechanics.

3. Minimize environmental footprint
   a. The extraction of mineral resources and the consequential need for the disposal of wastes, slurry, and water can have major environmental implications. Mining and mineral processing activities can generate massive amounts of toxic, corrosive, or flammable material. If released to the environment, these contaminants can have major impacts on surface water, groundwater, air, and land resources (NAE, 2010). The reduction of environmental impacts can be influenced by public expectations, best management practices, new legislation, and improvements in technology. Improvements in technology needed to decrease the impact of mineral processing on the environment include (NAE, 2010):
      - Reduce high energy consumption for grinding and slurry transport
      - Develop separation techniques that use fewer chemical reagents
      - Develop chemical reagents that are less toxic and more environmentally friendly
      - Develop environmentally acceptable techniques for disposal of tailings that usually contain toxic reagents
      - Develop models to predict the separation efficiency, economic effectiveness, and environmental implications of valuable material from non-valuable material
   b. According to the United Nations Berlin II Guidelines for Mining and Sustainable Development (2002), minimizing environmental impacts should rely on "sound management practices developed within a framework of good environmental legislation." Management practices such as incorporating new environmentally friendly technologies into mining processes, developing risk management plans, and educating workers on the linkage between the environment, ecology, and human health and safety can help to reduce environmental impacts.

4. Protect workers and the general public
   a. All industry has the responsibility to protect the health and safety of the public. Protecting public safety is especially important in the

mining industry as nearby populations can be negatively impacted by the release of particulate emissions, noise pollution, and mining waste products which can cause surface and groundwater contamination. Furthermore, mine employees are frequently exposed to high-risk work environments. Safety cultures need to be fostered in companies and should include everyone from management, to engineers, to operators. Training and education, high safety expectations for design and operation, and the development of both prescriptive and risk-based regulations can help prevent accidents.

1.3 MINING INDUSTRY TRENDS

To effectively address the challenges listed above, current mining trends need to be taken into consideration. With few exceptions, global production of mineral products has significantly increased in recent years (Table 1.1). This has not only led to the processing of larger volumes of materials, but also to the greater chance of negative environmental impacts. Trends in the mining industry can be used to predict which technologies are becoming cost-effective and where mining growth is expected in the future.

Population growth and urbanization are important factors that have, and will continue to, increase the demand for mineral commodities and in turn drive mining activities. According to the International Council on Mining & Metals (2012), the increase in urbanization and population in Asia is one of the driving forces in the recent increase in demand for mineral resources. Other current and future trends in the industry have been analyzed by ICMM (2012) and include:

- Shift in mining locations from developed to developing countries is a trend started in the mid-20th century
  - Growth of mining locations in Latin America, Africa, and parts of Asia is increasing.
  - While mining locations have shifted from developed to developing countries, mineral refinement has remained in developed countries.
  - Artisanal and small-scale mining accounts for a significant amount of world total production of tantalum, tin, and gold.
  - Young talent tends to be trained in one culture and language while available work is in another culture and/or language.
  - Major mining companies will come from China, India, and other developing countries in the future.

Table 1.1 Global production for select mineral commodities in 2010 and 2014 (U.S. Geological Survey (USGS), 2012, 2015)

Mineral/Product	2010 production	2014 production	Mineral/Product	2010 production	2014 production
Aluminum	40.8 Mt	49.3 Mt	Lead	4.14 Mt	5.46 Mt
Antimony	0.167 Mt	0.160 Mt	Lime	311 Mt	350 Mt
Arsenic	52,800 t	46,000 t	Magnesium compounds	5.76 Mt	6.97 Mt
Asbestos	2.01 Mt	1.98 Mt	Manganese	13.9 Mt	18 Mt
Barite	7.85 Mt	9.26 Mt	Mercury	2250 t	1870 t
Bauxite and Alumina	209 Mt	234 Mt	Mica	1.07 Mt	1.13 Mt
Bentonite	10.6 Mt	12.2 Mt	Molybdenum	0.242 Mt	0.266 Mt
Beryllium	205 t	270 t	Nickel	1.59 Mt	2.4 Mt
Boron	4.08 Mt	3.72 Mt	Peat	23.4 Mt	27.7 Mt
Bromine	0.45 Mt	0.411 Mt	Perlite	1.67 Mt	2.09 Mt
Cadmium	21,100 t	22,200 t	Phosphate rock	181 Mt	220 Mt
Cement	3310 Mt	4180 Mt	Platinum group metals	192,000 kg	161,000 kg
Chromium	23.7 Mt	29 Mt	Potash	33.7 Mt	35 Mt
Cobalt	89,500 t	0.122 Mt	Salt	280 Mt	269 Mt
Copper	15.9 Mt	18.7 Mt	Sand and gravel (industrial)	121 Mt	165 Mt
Feldspar	20.6 Mt	21.5 Mt	Silver	23,100 t	26,100 t
Fluorspar	6.01 Mt	6.85 Mt	Sulfur	68.1 Mt	72.4 Mt
Fuller's earth	3.29 Mt	3 Mt	Tin	0.265 Mt	0.296 Mt
Gold	2560 t	2860 t	Titanium	0.137 Mt	0.192 Mt
Graphite	0.925 Mt	1.17 Mt	Tungsten	68,800 t	82,400 t
Gypsum	147 Mt	246 Mt	Vanadium	57,600 t	78,000 t
Iron ore	2590 Mt	3220 Mt	Vermiculite	0.536 Mt	0.4 Mt
Kaolinite	33.1 Mt	41 Mt	Zinc	12 Mt	13.3 Mt

All values are in metric tons.

- There is an increasing need for talent that is technically excellent with high social skills to better relationships between host communities and countries.
- Current and future demand
  - Metal mining is currently dominated by iron, copper, and gold.
  - Future mining demand will be similar to that of today: coal, iron, gold, copper, bauxite, phosphate, potash, nickel, lead, and zinc.
  - Growth in demand originates for emerging economies that seek better material standards of living.
  - Only worldwide economic slowdown or disasters will slow the overall growth in mineral demand.
- Price of metals will remain high
  - Lag between demand and the rate at which supply can be increased via new mines will place upward pressure on prices.
  - Easily accessible and processable deposits are being depleted.
  - Expectations for environmental standards of performance are increasing.
  - New equipment, technology, processes, and associated training for staff requires capital.
- Shifts in technology
  - Mine production is shifting from underground mining to open-pit mining for some minerals.
  - Productivity increases in the past several years are due to the ability to process lower grade ores and through more efficient mineral processing.
  - Expanding exploration efforts and success in finding new deposits is the key to secure future metal supplies.
  - Continuous and automated mining operations are increasing.
- Estimated production trends
  - North America production will grow.
  - Latin America, Oceania (Australia and Papua New Guinea) will stay the same.
  - African production will grow.
  - Chinese production will grow, but not as rapidly as the past 10 years.
  - Areas of future growth: Siberia, Alaska, Northern Canada, Greenland, Nordic countries, and deep-sea mining.
- Company trends
  - More medium-sized companies will likely enter the market from emerging economies. There is currently a lack of medium sized

mining firms. As very large mining firms target deposits that will provide a mine lifespan of 20 years, medium-sized firms are needed to access deposits with shorter lifespans.

• Investor interest in corporate transparency will increase
• Companies that typically focus on smelting, refining, or other activities further down the value chain will become increasingly involved in extractive activities.

The NAE (2010) has stated that minimizing damage to the environment is a major challenge and is indeed the responsibility of industry and public agencies. Historically, industries have not felt a responsibility to minimize their environmental footprint, but this lack of accountability has been especially evident within the mining industry. Due to the pressure from regulatory agencies and the public, mining companies are now working in partnership with these groups to maximize national and community benefits while reducing environmental impacts and increasing safety (United Nations, 2002).

The top ten mining companies (as measured by revenue) in 2010 included BHP Billiton, Vale, Rio Tinto, China Shenhua, Xstrata (which has since merged with Glencore), Anglo American, Freeport McMoRan, Barrick Gold, Potash Corp, and Coal India. At least five of these companies (Coal India, Freeport McMoRan, Glencore, Vale, Anglo American daughter company AngloGold Ashanti) have been nominated for the world's worst company by the Public Eye in the past 10 years, demonstrating the need for improved environmental performance.

1.4 MINING CONTRIBUTION INDEX

It is difficult to enumerate the various ways mining affects a nation's economy as there are many direct and indirect factors impacting the mining industry's contribution to an economy. However, ICCM and Oxford Policy Management have developed a calculation that can be used to compare mining's effect on a country's economy (2014). This value is called the Mining Contribution Index (MCI), and is calculated using the following values from each individual country: the mineral export contribution in 2012, a change in the mineral export contribution from 2007–2012, and the 2012 production value as a percentage of the 2012 GDP. The MCI provides a reasonable estimate of the importance of mining and mineral processing for a country's economy; however, there are many other factors in the mining industry that could impact a country's economy (International Council on Mining & Metals (ICMM) et al., 2014). It is important to note that MCI does not provide indication of a country's contribution to global mineral consumption or production. The MCI was calculated for 2012 according to the following steps (International Council on Mining & Metals (ICMM) et al., 2014):

1. Countries were ranked in descending order according to three categories: mineral export contribution in 2012 as a percent of total merchandise exports, the change in the mineral export contribution from 2007–2012, and the 2012 total mineral production value as a percentage of the 2012 GDP. If no data exists for a given country, that country was omitted from the ranking of that category.

2. Percentile ranks were given to each country for each of the three categories based on the absolute country rankings. The absolute rank was divided by the maximum rank within that variable.

3. The three selection categories were weighed equally (1/3), summed, and multiplied by 100. A country that scored the highest in all three categories would have a MCI of 100.

4. For countries that did not have data for all three categories, the MCI was calculated giving equal weight (1/2) to the two available categories. Data were available for 201 countries for mineral export contribution and 198 countries for the change in the mineral export contribution from 2007–2012. Data were only available for 189 countries on 2012 production value as a percentage of the 2012 GDP.

For example, Australia is ranked 188 out of 201 and 178 out of 198 for mineral export contribution in 2012 and the change in the mineral export contribution from 2007-2012, respectively (higher ranks are better) (International Council on Mining & Metals (ICMM) et al., 2014). Australia is also ranked 160 out of 189 for 2012 production value as a percentage of the 2012 GDP. Thus Australia's MCI is 89.36, as calculated by the following formula (International Council on Mining & Metals (ICMM) et al., 2014):

$$MCI = \left(\frac{Absolute\ Rank_{Exp}}{Max\ Rank_{Exp}} + \frac{Absolute\ Rank_{\Delta Exp}}{Max\ Rank_{\Delta Exp}} + \frac{Absolute\ Rank_{\%GDP}}{Max\ Rank_{\%GDP}} \right) \left(\frac{1}{3} \right) (100)$$

where:

Absolute Rank$_{Exp}$ = absolute rank for mineral export contribution in 2010

Max Rank$_{Exp}$ = max rank for mineral export contribution in 2010

Absolute Rank$_{\Delta Exp}$ = absolute rank for the change in mineral export contribution from 2007–2012

Max Rank$_{\Delta Exp}$ = max rank for the change in mineral export contribution from 2007–2012

Absolute Rank$_{\% GDP}$ = absolute rank for the 2012 production value as a percentage of 2012 GDP

Max Rank$_{\% GDP}$ = max rank for the 2012 production value as a percentage of 2012 GDP

For countries without data for 2012 production value as a percentage of the 2012 GDP, the formula for calculating MCI is as follows (International Council on Mining & Metals (ICMM) et al., 2014):

$$MCI = \left(\frac{Absolute\ Rank_{Exp}}{Max\ Rank_{Exp}} + \frac{Absolute\ Rank_{\Delta Exp}}{Max\ Rank_{\Delta Exp}} \right) \left(\frac{1}{2} \right) (100)$$

The countries listed in Table 1.2 represent the countries that had the largest production value in 2012. The Human Development Index is a composite statistic that accounts for life expectancy, income, education, and standards of living. Mining and mineral processing contributes, on average, between 3–10% of a nation's GDP (International Council on Mining & Metals (ICMM) et al., 2014) although the percent of GDP for the largest producers ranges anywhere from 0.8–17.1%. In contrast, developing countries tend to have a higher average production value as percent of GDP (as high as 57%), indicating that certain low-income countries are reliant on their mining industry. This dependence does not necessarily have a negative impact on a country's economy. For example, some countries, such as Brazil and Chile, have realized a decrease in poverty compared to similar non-mining areas. Typically, smaller countries with lower incomes have high MCI values due to the method used for calculating the index. However, the countries with the largest production values may have relatively low MCI values due to having larger, more diverse economies that do not rely as heavily on mining. For example, Chile and Peru, while two of the larger producers in the world, are ranked 50th and 54th, respectively, according to MCI values on an overall global scale. The United States and China are ranked 67th and 130th according to MCI values, respectively (International Council on Mining & Metals (ICMM) et al., 2014).

Table 1.2 Mineral export and production data and mining contribution index values for select countries

	Total mineral export contribution 2012 (%)	2012 production value (US$ billion)	2012 production value as % of 2012 GDP	Human development index	Mining contribution index
Australia	69.0	153.23	10.0	0.93	89.36
Indonesia	40.7	54.96	6.3	0.68	78.46
Brazil	28.2	66.50	3.0	0.74	75.31
Canada	37.0	39.62	2.2	0.90	67.03
Chile	62.5	42.07	15.8	0.82	66.38
Peru	71.7	25.09	13.0	0.74	65.35
South Africa	41.7	65.46	17.1	0.66	62.43
United States	15.6	137.52	0.8	0.91	60.88
Kazakhstan	84.0	24.61	12.1	0.76	58.39
Russian Federation	77.3	89.77	4.4	0.78	54.39
India	30.1	89.69	4.8	0.59	51.20
China	2.9	501.98	6.1	0.72	39.54

Countries are listed in descending order according to MCI.
Data from International Council on Mining & Metals (ICMM) et al. (2014).

1.5 MINING PROCESSES

In order to understand how mining impacts the environment, it is important to understand the basic processes involved in mining and mineral processing, and at what phases environmental impacts can occur. Understanding mining processes is also vital during environmental assessment and when implementing an environmental management system.

Typical mining processes will vary depending on the target mineral or ore. Surface mining involves land clearing and overburden removal before minerals are exposed for excavation (Figure 1.1). Throughout all phases of surface mining, air pollution from dust and particulates is of primary concern. Underground mining involves excavating access shafts or ramps to the mineral deposit and the installation of ground support to maintain stability.

Once ore is excavated, it is crushed and ground to smaller sizes before going through concentration processes to obtain an ore or mineral concentrate (Figure 1.2). Left over materials and waste rock are disposed of as tailings. Crushing operations can contribute to air pollution, while crushing and separation methods using water and chemicals are a source of possible water contamination. Mine tailings also pose a risk of acid mine drainage and subsequent water contamination.

Figure 1.1 Simplified flow chart of a surface mining operation.

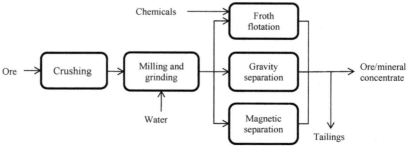

Figure 1.2 Simplified flow chart of ore/mineral processing. *(Adapted from Lottermoser (2010).)*

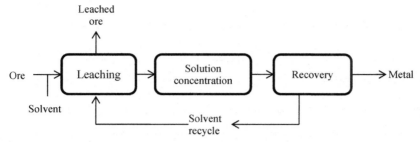

Figure 1.3 Simplified flow chart of hydrometallurgical processing. *(Adapted from Lottermoser (2010).)*

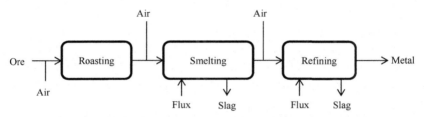

Figure 1.4 Simplified flow chart of pyrometallurgical processing. *(Adapted from Lottermoser (2010).)*

Once separated, ore concentrates can be further processed using hydrometallurgical (Figure 1.3), pyrometallurgical (Figure 1.4), or electrometallurgical processes to produce purified metals. Pyrometallurgical processes involve intense heat and can emit air pollutants every step of the way. Hydrometallurigcal processes can emit solvents during leaching and recovery operations and the water used throughout the process will contain chemicals that can contribute to water pollution.

1.6 OBJECTIVES

Mining and mineral processing are important activities for the industrialization of a nation and play a key role in the global economy. These activities are necessary for economic development, energy production, and international competitiveness. However, mining and mineral processing activities also have significant environmental impacts. Protecting the environment is a priority for all members of a society. Governments play a key role in setting environmental standards and ensuring individual and industry compliance. Increasingly, governmental agencies, industry, and community organizations are working together as partners to protect the environment for present and future generations. Developing mining and mineral processing, while minimizing environmental impacts, poses a significant number of challenges. This book focuses on such challenges.

Chapter 2 focuses on the environmental impact assessment of mining and mineral processing projects and provides a systematic step-by-step approach, along with proper guidance and considerations, for the procedure. It is important to note that effective environmental assessment in the early planning stages of mining activities, environmental management, and monitoring are vital to reducing long-term economic costs of environmental degradation and can contribute to the sustainable economic development of a nation.

Chapter 3 covers the implementation of crucial environmental management systems in the mining industry. Clearly, environmental management systems are important for lowering the risk of non-compliance, improving liability management and socioeconomic outcomes, and reducing long-term, operational, and closure costs. This chapter also covers costs and benefits of implementing environmental management systems and provides detailed descriptions of the steps required for successful implementation.

Direct, secondary, and long-term environmental and human health impacts from mining and mineral processing activities can be significant. Specific environmental impacts related to mining are described in **Chapter 4** and include the following:
- Air quality
- Water quality
- Acid mine drainage
- Ecological impact
- Land impacts
- Economic impacts

In order to successfully manage environmental impacts, monitoring is necessary to quantify impacts and determine the effectiveness of control measures. **Chapter 5** discusses monitoring implementation, procedures, technology, and data management.

With the implementation of an environmental management system, environmental auditing is necessary to ensure all environmental aspects are in compliance and continual improvements are being implemented. **Chapter 6** covers important aspects of environmental auditing and provides information regarding the steps necessary for a successful audit.

A thorough understanding of environmental auditing will facilitate the process and increase the likelihood of meeting audit criteria.

Chapter 7 contains information regarding control equipment used for environmental protection in mining and mineral processing. Control equipment is not only important for reducing environmental impacts, but can reduce worker exposure to potentially fatal or harmful pollutants.

Appendix A contains general air pollutant emission factors that can be used to calculate air emissions for various mining and mineral processing activities. For instances when site specific process emission factors are unavailable, general emission factors can be used to determine potential air quality impacts. General emission factors come from the U.S. EPA, the United Nations, and various other governmental and international agencies.

REFERENCES

International Council on Mining & Metals (ICMM), 2012. Trends in the Mining and Metals Industry: Mining's Contribution to Sustainable Development. Retrieved from: http://www.icmm.com/document/4441.

International Council on Mining & Metals (ICMM), 2013. 10 Principles. http://www.icmm.com/our-work/sustainable-development-framework/10-principles.

International Council on Mining & Metals (ICMM), Oxford Policy Management, Raw Materials Group, 2014. The Role of Mining in National Economies: Mining's Contribution to Sustainable Development, second ed. Retrieved from: http://www.icmm.com/document/8264.

Lottermoser, B.G., 2010. Mine Wastes: Characterization, Treatment and Environmental Impacts, third ed. Springer, New York.

National Academy of Engineering (NAE), 2010. Grand Challenges for Earth Resources Engineering. National Academy of Engineering. Retrieved from: https://www.nae.edu/File.aspx?id=106323.

U.S. Geological Survey (USGS), 2012. Mineral Commodity Summaries 2012. Retrieved from: http://minerals.usgs.gov/minerals/pubs/mcs/2012/mcs2012.pdf.

U.S. Geological Survey (USGS), 2015. Mineral Commodity Summaries 2015. Retrieved from: http://minerals.usgs.gov/minerals/pubs/mcs/2015/mcs2015.pdf.

United Nations, 2002. Berlin II Guidelines for Mining and Sustainable Development. Retrieved from: http://commdev.org/userfiles/files/903_file_Berlin_II_Guidelines.pdf.

World Coal Association, 2014. Sustainable Mining Practice Policy Statement. Retrieved from: http://www.worldcoal.org/bin/pdf/original_pdf_file/wca_sustainable_mining_policy_statement2014(29_07_2014).pdf.

World Commission on Environment and Development, 1987. Our Common Future. United Nations. Retrieved from: http://www.un-documents.net/our-common-future.pdf.

World Economic Forum Mining & Metals Industry Partnership, Accenture, 2014. Scoping Paper: Mining and Metals in a Sustainable World. World Economic Forum. Retrieved from: http://www3.weforum.org/docs/WEF_MM_MiningMetalSustainableWorld_ScopingPaper_2014.pdf.

CHAPTER 2

A Systematic Procedure for Environmental Impact Analysis of Mining and Mineral Processing Projects

2.1 INTRODUCTION

Most countries throughout the world require some form of environmental impact assessment (EIA) before approving new mining projects. Although regulations regarding EIAs vary from country to country, in all cases EIAs represent an important step to define and analyze environmental risks associated with a proposed project, identify ways to mitigate negative impacts, and are vital for sound and defensible decision making. Environmental impact assessments are meant to be used by decision makers and federal agencies to guide and plan further actions and inform the public of potential negative environmental impacts (Environmental Law Alliance Worldwide, 2010). Although an EIA may inform of environmental risks associated with a proposed project, there is no guarantee, in many places, that an environmentally harmful alternative will not be chosen (Environmental Law Alliance Worldwide, 2010). Rather, the EIA only ensures that any decision made is an informed decision. However, in some countries, local or federal regulations require the implementation of the least environmentally damaging practical alternative.

In the United States, the EIA process consists of an environmental assessment (EA) and (if necessary) an environmental impact statement (EIS). The National Environmental Policy Act (Council on Environmental Quality, 1978) defines an EA as a document that:
- Provides evidence and analysis to determine the need for an EIS or a Finding of No Significant Impact (FONSI)
- Identifies better alternatives and mitigation measures to reduce environmental effects
- Facilitates EIS preparation should one be required

Environmental Impact of Mining and Mineral Processing
ISBN 978-0-12-804040-9
http://dx.doi.org/10.1016/B978-0-12-804040-9.00002-4
17

Environmental assessments are concise and brief, providing information such as the need for the proposal, alternatives, environmental impacts, and agencies and people consulted. Environmental assessments conclude by reaching a Finding of No Significant Impact (FONSI) or determining that an EIS is required (Council on Environmental Quality, 1978).

Environmental impact statements are more rigorous and thorough than an EA. Environmental impact statements present full analyses and discussion for cases where significant environmental impacts are expected (Council on Environmental Quality, 1978). Thorough discussion is required for alternatives, including a "no-action" alternative, and the EIS must be released for public comment. The ultimate goal of the EIS is to insure that goals and policies of the National Environmental Policy Act are met.

This chapter presents a systematic procedure for preparing environmental documentation for mining and mineral processing projects. The major steps are presented in a sequential fashion, using hypothetical examples. Many different methodologies for impact assessment exist, both in theory and in practice. The provided procedure represents one simplified, systematic approach to the development of an environmental impact analysis for mining; using a matrix approach that relates a list of project activities to affected environmental attributes. Although the matrix method is used in this example, similar procedures using environmental attributes could be developed, and the reader's attention is directed to those places where variation is normally found.

2.2 SYSTEMATIC PROCEDURE FOR PREPARING ENVIRONMENTAL ASSESSMENT DOCUMENTATION

A step-by-step systematic procedure for preparing environmental assessment documentation is presented as a simplified flowchart (Figure 2.1). The degree of consideration to be exercised within some of the steps may vary with project scope and magnitude, but the basic algorithm is applicable in all cases. This systematic procedure is derived from the Handbook of Environmental Engineering Assessment (Jain et al., 2012). The procedure is based upon experience and processes used in the United States and many other industrialized countries. The concepts presented here are modified so that they are applicable to mining and mineral processing activities in China and other industrializing countries depending on specific regulatory, social, and economic needs.

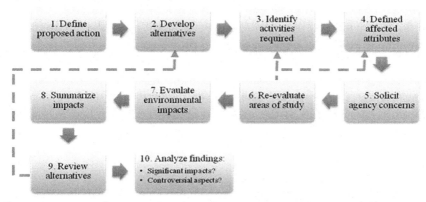

Figure 2.1 A step-by-step procedure for preparing environmental documentation. *(Adapted from Jain et al. (2012).)*

Define the Action

The first step in the environmental assessment process is to determine what
the proposed action is, and identify the purpose and need that it addresses.
At first glance, this step may seem unnecessary; however, this is often the
most difficult step of the process and may mean the difference between
agency success and failure if a court or administrative law judge reviews the
assessment. If the proponent is a typical government agency, a "proposed
action" may begin as an idea for change. Often these "good ideas" are
rather nebulous in nature, at least at first. An agency manager may want to
build a new mine or expand an existing mine, but may not have a thorough
understanding of why this would be a good idea— in other words, what is
the underlying purpose and need for the action?

Mining employees and staff may accept "because the governmental agency
wants to get this done" as a suitable reason, but this approach will rarely stand up
to the test of court, or public scrutiny. The environmental assessor must define
the underlying reasons why a new mine or proposed expansion would be
needed. Is a certain mining sector growing and requiries more resources?
Is an old mine no longer cost effective? Is an old mine ill-suited to the use of
modern technology? By answering these types of questions, assessors delve
into the purpose and need for the proposed action, which will in turn affect
what alternative courses of action are considered.

An environmental analysis relates to the physical actions proposed to
take place, and how they would affect the human environment. Even an
environmental impact assessment prepared on a new agency policy or

proposed legislation must be couched in terms of the physical actions that would be expected to ensue and how these might affect the environment. Although an environmental assessment is most useful if applied early in the planning process for a new proposal, meaningful assessment is not possible until the agency can identify firm descriptions of the physical actions expected under the proposal. The project description should be detailed enough to where each of the proposed actions can be linked to possible environmental impacts (Environmental Law Alliance Worldwide, 2010) and should include information on:

- Location
- Physical dimensions and footprint
- Construction timeline
- Possible materials and sourcing
- Operations and maintenance activities
- Treatment and disposal of wastes
- Environmental impact mitigation measures

Develop Alternatives

The development of alternatives, including the proposed action, is considered to be "the heart" of the environmental analysis. Why is this so important? At times it seems as if the engineering or economic studies prepared by the agency "prove" that there is only one "best" way to proceed with a proposed action, in part because these types of studies are generally focused on optimizing the agency's preferred approach rather than exploring options. In fact, every agency is aiming for that "best" plan. Regulations for environmental assessments specifically require that an agency considers alternative ways to reach a desired goal, and this is done to ensure that the agency does not foreclose reasonable options too early in the planning process. Alternative approaches may mean that the agency considers:

1. Different sites for a proposed action (e.g., the location of a new mine or the placement of a tailings dam)
2. Different designs for a proposed project
3. Seasonal timing for construction and/or phasing alternatives
4. Alternative construction/decommissioning techniques
5. Project size
6. Alternative operation and maintenance approaches
7. An alternative approach to meet agency needs without new construction

Several approaches are taken to ensure that the agency will consider a reasonable range of alternatives for meeting its underlying purpose and need for action, and disclose to the interested public and government agencies the full range of options under consideration. Additionally, a full range of alternatives will help the agency make better decisions and identify the most environmentally friendly options. However, it is important to remember that the "best" alternative may be one that can be implemented at reasonable cost without unreasonable delays, but may not always be the most environmentally friendly option. Agency goals may be better served in this manner than by fighting for many years to carry out the technically superior choice.

The agency should identify its "preferred alternative" as soon as is reasonable and at latest by the time that the final environmental assessment is circulated (but before a final agency decision has been made). If the agency has invested in engineering or economic studies, these may serve to support the agency's identification of its preferred course of action. Early identification of a preferred alternative will facilitate public disclosure and involvement in the process. An agency may wish to identify a preferred course of action in the beginning of the environmental review process; however, regulations caution against going too far and committing too many resources to an engineering solution until the environmental review is complete. In other situations, the agency may not be in a position to identify its preferred course of action until it has identified environmental impacts and considered public input through the environmental analysis process; in such cases, the preferred alternative may first be identified in the final EA.

Just as the definition of the proposed action requires some thought (see Step 1), the identification of alternatives also requires that the assessor think broadly to identify alternative ways of meeting the agency's stated purpose and need to take the action. The agency must accurately identify the need to which it is responding, or else the alternatives will not be responsive to the appropriate issue. For example, a mine operator may wish to build a dam. If the underlying purpose of this proposal is to control acid mine drainage (AMD) with a tailings dam, the alternatives considered by the agency will be very different from those developed if the underlying purpose of the dam is to create a reservoir for drinking water or even as a recycled source of water for mineral processing. In all cases, the proposed action might be essentially the same, but without knowing what is at issue, the mine planners may look at an inappropriate suite of alternatives (e.g., building a tailings dam to control AMD may require the dam to be located below gradient of the mine, while a dam for drinking water may require the dam to be up gradient of the mining operation). The Environmental Law Alliance Worldwide (2010) suggests that, at a minimum, an EA for a general mining process contains the following:

- Analysis of project alternatives
- Analysis of a "no-project" alternative
- Analysis of least environmentally damaging practical locations for mine facilities (e.g., tailings disposal, waste rock piles, leach facilities, etc.)
- Analysis of the least environmentally damaging practical alternative for ore extraction
- Analysis of least environmentally damaging practical ore beneficiation method
- Analysis of methods for disposal of waste and tailings

Alternative locations for mining facilities may not always be practical, since the location is highly dependent on ore deposits. However, depending on the geology, an alternative could be to switch from a surface mine to an underground mine, or to locate processing facilities and waste facilities to minimize impact to surface water or critical habitat (Environmental Law Alliance Worldwide, 2010). Alternative ore beneficiation (concentration) processes may, depending on the type of ore mined, include gravity concentration, milling and floating, leaching, dump leaching, or magnetic separation (Environmental Law Alliance Worldwide, 2010). Managing and disposal of mine tailings may typically involve either the release of tailings into the ocean, creating a tailings pond, or dry tailings disposal as paste backfill (Environmental Law Alliance Worldwide, 2010). Consideration should be given for the necessity, effectiveness, and environmental impacts of each alternative.

Must an environmental assessment reviewer consider endless arrays of possible alternatives? No. The agency must consider a "reasonable" range of alternatives. As this term is somewhat subjective, the practitioner may make use of the public and agency scoping process (Steps 5 and 6, Figure 2.1) to help decide what is a reasonable range. The agency may identify and reject alternative approaches that are very similar to alternatives that will be analyzed, alternatives that are not considered to be reasonable (such as those that are too expensive, or where necessary technology is not fully developed), or alternatives that are not responsive to the agency's stated purpose and need for the action. Remember that the analysis of alternatives is a way to help the decision maker consider the environmental consequences of the proposal and alternative approaches; consideration of very similar alternatives, or a large array of alternatives, may not be useful to help focus on those matters that are "ripe for decision." In practice, most environmental impact assessments consider between three and six alternatives, and most environmental assessments consider somewhat fewer. Agencies are expected to flesh out and analyze all alternatives in an EA to a "comparable" degree.

What is a practical alternative? This is a tough call. Broadly, a practical alternative accomplishes most or all of the agency's purposes. It should not be a frivolous action, which would serve no purpose if implemented, or an action that runs counter to law or societal needs, which could not be implemented. However, in the interests of public disclosure and edification, an agency may wish to describe "alternatives considered but not analyzed" or "alternatives dismissed from consideration" with a brief explanation of why these were not included in the full analysis (e.g., "this approach would require use of technology that is still in the experimental stage").

An environmental assessment analysis is a comparative analysis: the environmental impacts of taking an action are compared to the impacts of the different alternatives analyzed. An impact can be thought of as the degree of change that would be expected to occur over time. In order to determine the degree of change, the assessor must first determine what the baseline condition is; in other words, what would happen to the ambient environment if the action were not taken? The description of the environmental conditions that would be expected over time if no action were taken is often referred to as the "no-action" alternative. In many countries throughout the world, laws and regulations require that the effects of not taking the proposed action (or any action) are considered alongside the impacts of the proposed action and the alternatives analyzed (Environmental Law Alliance Worldwide, 2010).

In some cases, an agency may want to quit performing an ongoing activity. For example, a mining company may want to cease operation of a mine or of a certain mining technique. Some agencies take the position that ceasing an activity is not subject to an environmental assessment review; however, taking action to restore the site, such as decommissioning or tearing down a facility, or cleanup of stockpiles and mining waste, would be subject to review.

Identify Relevant Project Activities

In order to conduct an impact assessment, the assessor must know what activities would be expected to occur. To identify detailed activities associated with implementing the project or the program, agency activities may be categorized into functional areas. For each functional area, detailed activities associated with implementing projects or programs may be developed. The user should supplement these activities with project-specific activities. An example list of mining activities presented in Table 2.1 may be used as a starting point for analysis. Activities not applicable to the project should be crossed off, and supplemental activities should be added to encompass the project-specific requirements. If the agency has experienced construction project managers, they may be able to describe the phases of a typical agency project in some detail, and the assessor may build a similar table of relevant activities. Remember that terminology such as "Begin Phase II" is not inherently meaningful in an environmental assessment, but "clear the site" and "start mine excavation" can be related to real environmental effects.

Examine Attributes Likely To Be Affected

The environmental assessor should become familiar with the general nature of the individual attributes and the kinds of activities that may have an impact on them. A list of affected attributes may include (Environmental Law Alliance Worldwide, 2010; Jain et al., 2012):

- Atmospheric emissions
  - Methane
  - Particulate matter (dust)
  - Hazardous materials
- Ecological effects
  - Native flora and fauna
  - Habitat loss
- Land disturbance
  - Mine subsidence
  - Erosion
  - Land use patterns
- Water quality and quantity impacts
  - Surface water
  - Ground water
  - Acid mine drainage (AMD)

Table 2.1 Typical functional areas and associated mining activities

Exploration
- Earthmoving
- Clearing
- Grading
- Excavation
- Drilling
- Monitoring

Construction
- Clearing
- Grading
- Excavation
- Earthmoving
- Shaft/drift access and ventilation development
- Transportation and mobilization
- Construction of:
  - Building and conveyance system
  - Access roads and airstrips
  - Temporary or long-term living camps
  - Power supply (electricity, gas, or diesel)
  - Fuel and chemical storage facilities
  - Water supply
  - Process plant
  - Workshops and warehousing
  - Contractor lay down areas
  - Offices, change rooms
  - Crushing plant
  - Tailings storage facilities
  - Waste rock, low-grade and other dumps
  - Stockpile preparation

Operations
- Drilling and blasting
- Removing/moving/storing overburden
- Extracting/transporting/dumping ore
- Crushing materials, grinding circuits
- Screening
- Storage/disposal of tailings
- Washery (water used to remove dirt from a mineral, especially coal)
- Mineral processing, beneficiation of material
- Transporting and placing washery rejects

Operations, Cont'd.
- Workshop and/or power plant operations
- Creation of pit/stockpiles/ exposed areas
- Transportation: rail transport, ship loading, activities at ports
- Equipment: dozers, excavators, loaders, haul trucks, conveyors, drills, blasting, processing plants
- Materials being tipped and separated
- Conveyors, rill towers, rail/ truck unloading facilities
- Waste rock piles, ore stockpiles, tailings storage facilities/dams, heap/dump leach pile
- Water management, e.g., watering down stockpiles
- Planned subsidence

Closure/Rehabilitation
- Demolition and removal of infrastructure
- Landform reconstruction: re-grading, covers, slopes
  - Creation of covers for tailings/stockpiles
- Fencing off, sealing, or filling in voids or open pits
- Completing the rehabilitation and remediation processes
  - Re-vegetation
  - Establishment of native species/animals
- Monitoring and measuring the performance of closure activities against the agreed standards and criteria
  - Monitoring for acid mine drainage
  - Inspections, consultation and reporting to stakeholders on progress
- Progressive community and government sign-off

- Human impacts
  - Human health
  - Public safety
  - Traffic
- Economic effects
- Social impacts
  - Changes in population characteristics
  - Changes in access to land, water, and basic services

See Chapter 4 for more detailed descriptions of the environmental attributes and descriptor packages that may be used to identify areas where available technical expertise is deficient and additional assistance may be required.

Soliciting Government, Industry, and Public Concerns

Throughout the environmental impact review process, the agency is expected to listen to, identify, and address environmental concerns raised by the government agencies or external parties, including industries and the public. In particular, an agency with regulatory expertise is a vitally important party. One of the purposes of environmental assessments is to open mining industry decision making to government agency scrutiny; a corollary to this is the challenge to encourage and facilitate public involvement in industry decisions. An agency can solicit government and public involvement at specific points when it prepares an EA.

Under the process for preparing an EA, regulations provide for government input at the following specific points:

- Public scoping process – After announcing its intent to prepare an EA, an agency must invite participation from affected federal, state, and local governments, other interested parties, and the general public to help the agency determine the scope of the analysis.
- Review and comment on draft environmental impact assessment – Before preparing a final EA, an agency must make a draft version available to the public for review and comment. The review and comment period must be at least 45 days long, but is often longer.
- Review and comment on final environmental assessment – An agency must make a final EA available to the government agency and public prior to taking any action on the proposed project. At the agency's discretion, it may solicit comments on a final EA prior to making a final decision.

As part of their own regulations or guidance on how to conduct the EA review process, many agencies have adopted detailed participation procedures and provisions for giving notice to government agencies and the public. Some agencies have statutory or administrative processes that must be considered alongside the environmental review process. An agency can go beyond the requirements of law or regulation and involve the public or other parties at additional points in the assessment process. For example, an agency may circulate a plan on how it will approach preparing the environmental review; seek government agency input into a participation plan; provide advance notice of its intent to prepare a future environmental review; conduct a pre-scoping exercise to determine if a suggested proposed action is warranted; circulate more than one draft assessment for review and comment; prepare and circulate ancillary documents such as a technology review of alternatives, a rationale for the preferred alternative, or an economic feasibility report; or circulate a draft decision document for review. At some point, however, the agency must forge ahead with the analysis in order to support a meaningful agency decision and action.

Re-Evaluate Identification of Activities and Attributes

After consulting with other agencies, interest groups, and the public, the proposed activities that will be required and attributes that may be affected should be re-examined. The dashed lines in Figure 2.1, which lead from Step 6 back to Steps 3 and 4, represent this re-examination and reformulation following the government involvement stage. The members of the various outside groups may have additional concerns that had not been addressed or may not be particularly concerned about some elements that the mining agency believed to be important. It is at this stage that the plans for the proposed action first may be modified to become more acceptable, or the planning for content of the EA may be redirected toward the elements of greatest concern.

The agency should not avoid mention of a known attribute just because the public is not concerned. There is still a responsibility to cover all relevant attributes. Instead, the primary focus should be placed on aspects that are of greatest concern to the public. If the comments are from a government agency with regulatory expertise, the comments should be given great deference.

Evaluate Impacts Using Descriptor Package and Worksheets

Using the activities developed in Step 3 and the attribute list, the assessor may find it useful to construct a matrix worksheet, with activities on one axis and environmental attributes on the other. Figure 2.2 indicates an example format. The attribute descriptor packages (see Chapter 4) may be used to identify environmental attributes.

The matrix in Figure 2.2 can be used to:
- Identify potential impacts on the environment by placing an "X" at the appropriate element of the worksheet
- Collect baseline data on the affected attributes
- Determine areas where energy savings could occur
- Quantify the impact where possible using an analytic approach

For instance, the construction of a surface mine will almost certainly require large-scale excavation, which might later cause erosion that could then result in increased suspended solids in the receiving waters of a nearby stream and cause a decrease in dissolved oxygen. An "X" may be marked on the worksheet for all potential negative impacts, and a "+" for any potential positive impacts. It should be emphasized that this evaluation is to be done on an interdisciplinary basis.

It is important to note that EAs should be analytic rather than encyclopedic. The purpose, of course, is to reduce the bulk of the EA, focus on the parameters of greatest interest, and make the document most useful to decision makers. It is therefore essential that the EA includes analytic and quantitative information for environmental impacts where possible. Lengthy subjective discussion of the impacts and general "boilerplating" of the document text should be avoided. For example, including standard lists of endangered species found somewhere in the province is of little value—include only discussion of those species known or thought to be present in the vicinity of the project during all or some portion of its duration. The decision maker needs to know if there are any endangered species or critical habitats on the site or in an area affected by the action but does not need to consider that another species might be present in a different type of habitat in a different area.

In the surface mining example, the analysis started with a consideration of the proposed action, that is, constructing a surface mine. However, in some cases, it may be more useful to start with an analysis of the existing case as described in the no-action alternative. This is applicable in situations where the agency is considering alternative approaches but

Figure 2.2 Example impact matrix for a mining project. Actual impacts will vary on a case-by-case basis. *(Adapted from Jain et al. (2012).)*

	Attributes				
	Air Quality	Water	Ecological Impacts	Land Impacts	Economic Impacts

Project / Activities:

Exploration
- Earthmoving
- Excavation
- Drilling
- Monitoring

Construction
- Clearing
- Grading
- Excavation
- Transportation
- Buildings/Facilities
- Storage
- Access roads
- Utilities

Operations
- Drilling/blasting
- Extracting
- Crushing and grinding
- Screening
- Washery
- Mineral processing
- Workshops
- Transportation
- Equipment
- Conveyors
- Waste storage and disposal
- Water management
- Subsidence

Closure
- Demolition
- Landform reconstruction
- Sealing and fencing
- Re-vegetation
- Monitoring

Attribute columns:

Air Quality: Diffusion factor, Particles, Sulfur oxides, Hydrocarbons, Nitrogen oxides, Carbon monoxide, Photochemical oxidants, Hazardous toxicants, Odor

Water: Aquifer safe yield, Flow variations, Dissolved oxygen, Nutrients, Suspended solids, Dissolved solids, Oil, Acid mine drainage, Radioactivity, Toxic compounds

Ecological Impacts: Terrestrial vegetation/fauna, Aquatic fauna, Aquatic, riparian, and groundwater dependent vegetation

Land Impacts: Erosion, Top soil loss, Subsidence, Land use patterns

Economic Impacts: Backward linkage, Forward linkage, Fiscal linkage, Consumption linkage has not yet formulated a proposed action or preferred course of action. Examples would be considerations of changes to existing mineral recycling practices and development of a large-scale resource management plan.

Summarize Impacts

For potential impacts marked with "X" or "+" on the worksheet (Figure 2.2), the results can be summarized using Figure 2.3. Areas of net positive or net negative impacts can be shaded in, using the shading color to indicate the significance of the impact (some practitioners use a red-yellow-green color shade for adverse impacts, areas of potential

Economic Impacts	Consumption linkage		30	
	Fiscal linkage		29	
	Forward linkage		28	
	Backward linkage		27	
Land Impacts	Land use patterns		26	
	Subsidence		25	
	Top Soil Loss		24	
	Erosion		23	
Ecological Impacts	Aquatic, riparian, and groundwater dependent vegetation		22	
	Aquatic fauna		21	
	Terrestrial vegetation/fauna		20	
Water	Toxic compounds		19	
	Radioactivity		18	
	Acid mine drainage		17	
	Oil		16	
	Dissolved solids		15	
	Suspended solids		14	
	Nutrients		13	
	Dissolved oxygen		12	
	Flow variations		11	
	Aquifer safe yield		10	
Air Quality	Odor		9	
	Hazardous toxicants		8	
	Photochemical toxicants		7	
	Carbon monoxide		6	
	Nitrogen oxides		5	
	Hydrocarbons		4	
	Sulfur oxides		3	
	Particles		2	
	Diffusion factor		1	
		Net Positive Impact +	Attribute Number	Net Negative Impact X

Figure 2.3 Example matrix for summarizing possible impacts for a mining project. Actual impacts will vary on a case-by-case basis. *(Adapted from Jain et al. (2012).)*

concern, and positive impacts). For example, for impacts on erosion, suspended solids, and dissolved oxygen, the magnitude of the project is evaluated along with site characteristics and scientific information to determine the degree of severity of the impact on each attribute. Finally, the impacts on each attribute from all project activities are summarized by using the key shown in Figure 2.3. In practice, most agencies concentrate on identifying and quantifying adverse impacts instead of positive impacts.

Review Alternatives

After summarizing the impacts from each alternative, it is important to review the results. It may be determined that the current alternatives are not satisfactory or that other previously overlooked alternatives may be practical. If so, new alternatives should be considered. The dashed line in Figure 2.1, which flows from Step 9 back to Step 2, represents this re-evaluation process and the steps for evaluating new alternatives should follow the same procedure as previously outlined. Examples of alternatives for the surface mining project example include:

- Alternatives related to different designs and/or projects, such as the utilization of different mining practices or mineral recovery
- Alternative measures to provide for mitigation of fish and wildlife losses discovered to be associated with soil erosion problem, or alternative construction methods to be more energy efficient
- Improving the efficiency of mineral recycling programs through new technology or more effective marketing strategies

Remember to be imaginative. It may be necessary to evaluate alternatives that are not within the prerogative of the proposing agency. This is specifically required if the alternative is otherwise reasonable.

Analyze Findings

The final step is to analyze the findings of the EA, determine whether significant adverse environmental impacts would be expected, and identify mitigation measures. Among other requirements, an environmental review is used to do two things: disclose to the government agencies, other industries, and the public whether the environmental impacts of a proposed action would be significant, and assist in developing mitigation measures to lessen adverse impacts. An EIS should be prepared for "major actions," defined as those actions having significant environmental impacts. To determine whether impacts should be considered significant, the following questions may provide guidance:

• Will the implementation of the action or program have a significant adverse effect on the quality of the human environment?
• Will the action be deemed environmentally controversial? (Controversy caused by other considerations, such as local politics or economics, does not, by itself, trigger this requirement)
• Are the possible adverse environmental impacts highly uncertain or do they involve unique or unknown risks? (In other words, is the assessor unable to determine if impacts would be significant or not?)

If the answer to any of the above three questions is "yes," the mining agency needs to prepare an EIS before executing the proposed activity. Similarly, if the impacts, while individually not significant, cross a threshold of significance cumulatively, an EIS would also be required. In some cases the mining agency may decide that it is preferable to abandon or delay the project until uncertainties can be dealt with or the project can be reconfigured rather than proceed with preparing an EIS. Additionally, some jurisdictions may have laws that require the implementation of the least environmentally damaging practical alternative, which is some cases may be the no-action alternative. If the answer to all of the above three questions is "no," then the agency can complete the EA with a finding of no significant impact (FONSI) and an EIS is not required and the project may proceed. Regardless, the findings must be made available to government agencies and the public.

Often, a look at the summary sheets from Step 8 will demonstrate that many impacts are the same for all alternatives. In this case, the assessor may document in the environmental review those impacts that are common to all alternatives (including mitigation measures that may be applied), and eliminate these from further discussion or analysis. This allows the agency to concentrate its environmental review on the impacts that vary by alternative, thereby assisting in bringing into sharp focus those environmental considerations that bear on the decision to be made.

A second use of the analysis of findings is identifying where mitigating measures would be useful in eliminating an adverse impact or reducing its degree of adversity. Some adverse impacts cannot be mitigated. Sometimes the mitigation itself might create problems for other aspects of the environment (e.g., mitigations to assist wildlife might have an adverse impact on cultural resources). Sometimes it is not clear that a suggested mitigation measure would substantially ameliorate the adverse impact. For these reasons, the mitigation measures, once identified, are then evaluated under

Step 7, and the benefits or detriments are captured in the summary sheets prepared under Step 8. Step 10 is then revisited to address the effect of the mitigation measures under each alternative.

It is important to note that mitigation has two different meanings in the environmental review process context. It may refer either to measures taken to avoid or lessen an adverse impact, or measures taken to compensate or "atone" for some potential damage or loss. The latter context is regularly seen when dealing with habitat loss, particularly of wetlands. Mitigation measures can be included in the EA. A post-project mitigation-monitoring plan may be required to ensure the promised mitigations are implemented.

When analyzing impacts and determining mitigation measures, it is important to identify any assumptions made. This ensures that both the casual reader and the professional reviewer can understand how the conclusions were reached. For example, if a proposed mine would be built in an area of moderate population growth, it makes a critical difference in total socioeconomic impacts if the analyzer assumes continuation of the same rate of moderate growth over a 20-year analysis period, or a much greater rate of growth (perhaps due to the proposed project or perhaps due to cumulative impacts) over the same period. Stating the analysis assumptions also helps to ensure that analyses of impacts on different types of attributes are consistent. In the example given, if the socio-economist assumes continuation of a moderate population growth with concomitant impacts on land use, the air-quality analyzer must assume the same growth rate when determining potential impacts on air quality. If one analyzer assumes a moderate growth rate while another assumes a high growth rate, the comparative impacts will be skewed. It is worth pointing out that stating obvious assumptions such as "all laws will be followed" is not particularly useful; no one would reasonably propose a project where laws would not be followed.

REFERENCES

Council on Environmental Quality, 1978. Regulations for Implementing the Procedural Provisions of the National Environmental Policy Act. CFR Parts 1500–1508. Retrieved from: https://ceq.doe.gov/ceq_regulations/Council_on_Environmental_Quality_Regulations.pdf.

Environmental Law Alliance Worldwide, 2010. Guidebook for Evaluating Mining Project EIAs, first ed. Retrieved from: http://www.elaw.org/files/mining-eia-guidebook/Full-Guidebook.pdf.

Jain, R., Urban, L., Stacey, G.S., Balbach, H., Webb, M.D., 2012. Handbook of Environmental Engineering Assessment: Strategy, Planning, and Management. Elsevier, Waltham, MA.

CHAPTER 3

Environmental Management System Implementation in the Mining Industry

3.1 INTRODUCTION

An environmental management system (EMS) is a component of a mining management system that pertains to the procedures, responsibilities, and processes for preventing harmful environmental, economic, and social impacts. An environmental management system helps the mining company achieve leading practices by providing a framework for continuous improvement (Australian Centre for Sustainable Mining Practices, 2011; Environment Australia, 2002; Hilson and Nayee, 2002; National Mining Association, 2012). In the past, economic and financial gains were heavily emphasized in the mining industry and little emphasis was placed on mitigating mining impacts on the environment, resulting in environmental degradation. Environmental degradation is no longer acceptable and balances between economic gains, local community benefits, and environmental impacts can be achieved to create advantages for all. To gain local acceptance, stakeholder backing, and regulatory approval, the mining industry needs to increase their responsibility to the environment and local communities. Environmental management systems are currently used to efficiently manage environmental impacts, ensure regulatory compliance, reduce costs and risks, and improve reputation, and meet stakeholder demands (National Mining Association, 2012).

To successfully implement and operate a mine in an environmentally, economically, and socially responsible manner, mine management must be aware of the following:
- What is an EMS?
- Costs and benefits of implementing an EMS and best practices (BPs)
- How to encourage and promote BPs throughout the mining industry
- Techniques to implement an EMS or improve an existing EMS
- The different management aspects of an EMS

Environmental Impact of Mining and Mineral Processing
ISBN 978-0-12-804040-9
http://dx.doi.org/10.1016/B978-0-12-804040-9.00003-6

3.2 WHAT IS AN EMS?

An environmental management system (EMS) is designed to help companies manage their environmental liabilities and responsibilities and involves a continuous cycle of improvement (Hilson and Nayee, 2002; National Mining Association, 2012; Newbold, 2006). The different components of an EMS include identifying areas and operations that require controls, establishing objectives/goals/standards, implementing process controls that will help meet those standards, monitoring, documenting, and auditing the processes, and reviewing the monitoring and auditing data to make improvements where needed. This is often summarized as a "plan-do-check-act" model of management. Standalone environmental auditing, while beneficial, is not sufficient to achieve the level of clean production needed to alleviate mining impacts on the environment. In fact, environmental monitoring, assessment, and measurements link audits, reviews, and documentation together, creating a environmental management system (Hilson and Nayee, 2002). Continuous improvement for decreasing mining impacts and increasing efficient use of resources becomes a responsibility inherent in mining processes, not an after thought. Figure 3.1 shows the cyclical process of EMS.

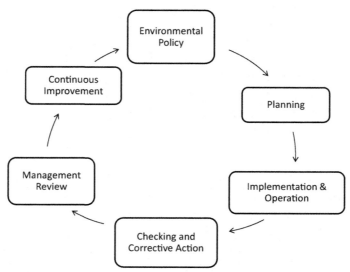

Figure 3.1 The EMS cycle. *(Adapted from Commission for Environmental Cooperation (CEC) (2005) and Hilson and Nayee (2002).)*

Tole and Koop (2013) further describe the major steps when adopting an EMS:

- Development of an environmental policy that is supported by senior managers
- Identification of legal/regulatory requirements and determining areas that require improvements to reduce environmental impacts
- Development of a system for implementation, assignment of responsibilities, creation of communication channels, and documentation of the EMS and procedures for control processes
- Development of a system for monitoring, measurement, and improvement (including reporting and non-compliance)
- Continual review by senior management of the effectiveness of the EMS to meet objectives and compliance goals

Certifications for EMSs exist to create a global standard for management, auditing, and monitoring procedures. The most well known certification for environemntal management is ISO 14001, the International Organization for Standardization (ISO) code that gives flexibility to the mining company to design their own EMS to cover 17 specific elements (International Organization for Standardization (ISO), 2013; National Mining Association, 2012):

- Environmental policy: Develop an environmental policy committed to prevention, improvement, and compliance.
- Environmental aspects: Conduct an EIA and identify areas that require controls or process improvements.
- Legal and other requirements: Determine legal requirements, stakeholder interests, and company requirements for the development of an EMS.
- Objectives, targets, and plans: Outline objectives. Develop and implement plans to meet target objectives based on regulations and requirements.
- Resources, roles, responsibility, and authority: Assign resources and personnel for EMS implementation. Document roles and responsibilities.
- Competence, training, and awareness: Implement and document training programs. Promote awareness of policy, procedures, and EMS requirements.
- Communication: Effective communication of goals and objectives from management to staff and communication of performance and monitoring from staff to management.
- Documentation: Develop proper documentation for procedures and main elements of the EMS.

- Document control: Develop document control system for proper access, version control, and protection from damage.
- Operational control: Management of operations via engineered controls or procedures to reduce environmental impacts. May incorporate procedures to manage operational changes.
- Emergency planning and response: Develop emergency response plans and resources. Provide emergency training, equipment, and continual adjustments.
- Monitoring and measurement: Measure, monitor, and document non-conformance events. Maintain monitoring equipment.
- Evaluation of compliance: Verify legal and regulatory compliance.
- Non-conformance, corrective and preventative action: Identify, investigate, and correct non-conformance incidents. Review the effectiveness of preventative and corrective actions.
- Record control: Maintain records of EMS performance and compliance.
- Internal audit: Verify the effectiveness of the EMS at meeting objectives and requirements. Provides audit results for review.
- Management review: Periodic review of potential improvements and taking action.

It is important to note that ISO 14001 can be applied to any type of organization, and while it lists the required elements of an EMS, it does not state how they should be implemented (National Mining Association, 2012). Thus, individual organizations can modify the model to meet their specific objectives and comply with local and regional regulations.

Receiving ISO certification, while encouraged, is dependent on the mining company. Benefits of receiving ISO certification include increased stakeholder acceptance and peace of mind for community members. On the other hand, certification can be expensive, take a while to implement, and is not necessary to ensure that appropriate procedures are being followed to manage the environmental components of the mining industry. Smaller mining companies may not always receive a certification due to budgetary restraints, but might use the ISO procedure as a guideline (Hilson and Nayee, 2002).

Increased stakeholder and regulatory acceptance may be the motivation for companies receiving ISO certification, not through a concern for the environment. An inherent issue with ISO certification is the lack of transparency required in third party auditing documents (Tole and Koop, 2013), resulting in little to no accountability. Due to misplaced motivations and this lack of transparency, ISO may not always be implemented effectively. In order to be effective, ISO and other certifications must be implemented with environmentally minded motivations.

3.3 BENEFITS OF IMPLEMENTING AN EMS AND BEST PRACTICES (BPs)

The driving ideas behind EMSs stem from sustainable development principles. As global trends lean towards more sustainable development, a change in business thinking and attitudes will be needed to incorporate sustainable development principles into EMSs within the mining industry.

Incorporating sustainable development ideals into EMSs and implementing BPs minimize adverse environmental, social, and economic impacts, whereby resulting in the following benefits (Australian Centre for Sustainable Mining Practices, 2011; Darnall and Pavlichev, 2004; International Atomic Energy Agency, 2010; Ireland Environmental Protection Agency, 2006; National Mining Association, 2012; U.S. Environmental Protection Agency (U.S. EPA), 2013):

- Improved liability management, environmental management, quality control, process efficiency, and socioeconomic outcomes
- Lower risk of non-compliance
- Reduced operational, long-term, closure, and rehabilitation costs
- Improved accountability and corporate governance
- Enhanced employee awareness of environmental issues and increased morale
- Greater acceptance from stakeholders, the community, investors, insurers, non-governmental organizations (NGOs), buyers, and suppliers

Environmental management systems can be implemented in various ways depending on the desired outcomes. If greater acceptance from the community and stakeholders is desired, an EMS can include programs for public outreach during project planning or include an environmental policy that all can agree upon. The early adoption of some components of an EMS, such as public outreach and cooperation can be vital during the initial steps of mine exploration and construction.

Case Study: Coal Mining and Aboriginal Peoples of British Columbia

Canada is a country rich in natural resources and is one of the top mineral producers in the world. Canada's mineral resources brought in an estimated

$43 billion in 2013, with coal accounting for $7 billion (Natural Resources Canada, 2014). A large portion of the coal mined in Canada is from British Columbia, Alberta, and Saskatchewan.

Coal deposits in the Mount Klappan area of British Columbia represented one of the world's largest undeveloped anthracite coal resources, with 108 million metric tons of coal measured, and a total of 2.2 billion metric tons classified as inferred or speculative (Fortune Minerals Ltd., 2009). Anthracite coal is of the highest rank (carbon content) and has the lowest moisture content of all coals, making it attractive for clean coal power generation and for use in the steel industry. Fortune Minerals Ltd., obtained the rights to the Mount Klappan area in 2002 from a large oil company, which had already conducted significant exploration and testing. However, the Mount Klappan site, consisting of 15,000 hectares, was located in the traditional territories of the Tahltan Nation, which holds the aboriginal title and rights to their land (Fortune Minerals Ltd., 2009).

Tensions mounted in 2005 as environmental and feasibility assessments were beginning at the proposed mining site. Protesters from the aboriginal tribes blockaded the main access road to the mining site as a result of the mining project's infringement on aboriginal titles and rights. Internal conflicts in the native communities led to agreements between Fortune Minerals and the natives' central council that were not supported by a portion of the community. Due to the lack of internal communication, many natives were not aware that mining and development agreements even existed. Protestors felt that because the project was located on their traditional lands and they would be the most affected as a result, direct engagement was appropriate. Protesters wanted not only to protect their native lifestyles, cultural sites, and territory, but they also wanted to protest unsustainable development practices within the industry and the infringement of their native rights to govern their traditional territories. The goals of the protestors were to stop resource development until agreements on sustainability, governance, shared-decision making, and revenue sharing could be reached with the mining company. Tensions between Fortune Minerals and the Tahltan Nation has remained high with ongoing blockades and protests, despite reaching an environmental assessment cooperation agreement in 2009, establishing open dialogue and mutual recognition of each other's interests (Fortune Minerals Ltd., 2009).

This situation demonstrates the need for an effective public outreach program and communication with the local communities during all steps of mining operations. The native communities were upset that their council approved of such operations without their consent. However, this situation could have been avoided by direct contact by the mining company with both the communities and the leadership, allowing community concerns to be conveyed without the need for drastic actions such as protesting and blockades. Although it is hard to say with certainty that such programs would have completely prevented this situation, it is unlikely that open communication and involvement of communities in the mining process would have further damaged relations. The cost of establishing a public outreach program is often modest, but protests from local communities can stop progress on major mining projects, threatening entire project timelines and budgets.

3.4 COSTS OF IMPLEMENTING AN EMS AND BPs

Because the shift toward sustainable practices has been so significant, the mining industry needs to approach the issue of cost from the point of view of how much it would cost **not** to implement an EMS (Holtom, 2010). A poorly implemented or complete lack of an EMS may scare away investors, reduce cooperation with stakeholders, and increase operational, closure, and remediation costs.

Implementing an EMS, along with BPs, requires a high initial capital costs, but result in large savings over the long run. The cost of integrating an environmental management system into all aspects of mining can be around 5% of capital and operating costs for new mines (Environment Australia, 2002). Despite this ballpark figure, the amount a mine will spend on the EMS and BPs differs depending on the location and existing economic and environmental conditions. Sturm and Upasena (1998) found that mining operations in industrialized countries tend to spend more on software, training, and environmental handbooks while mining in developing countries tend to spend more on hardware, such as pollution abatement equipment and recycling and waste management processes. Regardless, maintaining an effective EMS requires ongoing labor costs (and possibly outside consulting), which represent the majority of expenditure on EMSs (U.S. Environmental Protection Agency (U.S. EPA), 2013).

To achieve the greatest cost savings, developing best environmental practices during the mine planning phase is crucial. This will save money on retrofitting and redesigning systems later in the mine life. Savings can be seen in both the long term as well as short term. Short-term savings include savings due to better managed water and energy systems as well as the avoidance of unacceptable environmental impacts and any associated non-compliance fines

(Environment Australia, 2002). In the longer term, savings are seen through decreased future costs for mine decommissioning. Less rehabilitation is needed during mine closure because control processes have been incorporated from the beginning of the mine life and cleanup has been executed as the mine develops. This is more efficient and cost effective as equipment is still on site and cash flows still exist. For example, there are approximately 500,000 abandoned mines in the United States, with an estimated management and pollution remediation cost of $20 billion (Pacific Institute, 2013). Many of these mines are expected to require management in perpetuity. In addition to remediation, savings can also be seen through improved acceptability for future mining operations. Arguably, more forward thinking environmental management can also have a positive impact on mining safety, resulting in increased savings through reduced accident rates.

3.5 GOVERNMENT INVOLVEMENT IN EMS

Governments have a responsibility to reduce environmental degradation. There are several different ways government can be involved, including setting standards and environmental regulations (as well as EMS implementation), enforcing standards, and offering incentives for high performers or punishing low performers (Uchida and Ferraro, 2007). Past experiences in the United States have demonstrated that government involvement in increasing environmental standards and regulations with accompanying enforcement has resulted in reduced environmental degradation and increased adoption of EMSs (Darnall and Pavlichev, 2004). It is important to note that environmental goals and performance are not inherently dictated by an EMS, but are set by each individual company to meet their desired goals.

The role of government in environmental policy is often debated, though many see government involvement as a necessary force. Although other drivers exist, government regulation remains the primary motivator for adoption of EMSs (Darnall and Pavlichev, 2004). To promote effective policies, governments should have standards or principles to guide their decision making processes. The Berlin II Guidelines for Mining and Sustainable Development (United Nations, 2002) provide 15 fundamental principles; it is vital that they be followed by governments and industry:

1. Recognize environmental management as a high priority, notably during the licensing process and through the development and implementation of environmental management systems. These should include early and comprehensive environmental impact assessments, pollution

control and other preventive and mitigative measures, monitoring and auditing activities, and emergency response procedures.

2. Recognize the importance of socio-economic impact assessments and social planning in mining operations. Social-economic impacts should be taken into account at the earliest stages of project development. Gender issues should also be considered at a policy and project level.

3. Establish environmental accountability in industry and government at the highest management and policy-making levels.

4. Encourage employees at all levels to recognize their responsibility for environmental management and ensure that adequate resources, staff and requisite training are available to implement environmental plans.

5. Ensure the participation of and dialogue with the affected community and other directly interested parties on the environmental and social aspects of all phases of mining activities and include the full participation of women and other marginalized groups.

6. Adopt best practices to minimize environmental degradation, notably in the absence of specific environmental regulations.

7. Adopt environmentally sound technologies in all phases of mining activities and increase the emphasis on the transfer of appropriate technologies that mitigate environmental impacts, including those from small-scale mining operations.

8. Seek to provide additional funds and innovative financial arrangements to improve environmental performance of existing mining operations.

9. Adopt risk analysis and risk management in the development of regulation and in the design, operation, and decommissioning of mining activities, including the handling and disposal of hazardous mining and other wastes.

10. Reinforce the infrastructure, information systems service, training, and skills in environmental management in relation to mining activities.

11. Avoid the use of such environmental regulations that act as unnecessary barriers to trade and investment.

12. Recognize the linkages between ecology, socio-cultural conditions and human health and safety, the local community, and the natural environment.

13. Evaluate and adopt, wherever appropriate, economic and administrative instruments such as tax incentive policies to encourage the reduction of pollutant emissions and the introduction of innovative technology.

14. Explore the feasibility of reciprocal agreements to reduce transboundary pollution.

15. Encourage long-term mining investment by having clear environmental standards with stable and predictable environmental criteria and procedures.

Non-mining industries found that by taking a proactive approach to existing environmental policies and accepting the responsibility for self-assessment through developing EMSs, they were able to anticipate government involvement and policy and make changes to prevent issues from becoming challenges, thereby reducing government mandates (Newbold, 2006). As minimal government intervention is desired by the mining industry, the mining industry has attempted to replicate this effect. Indeed, studies in the 1990s found that after initial governmental pressure was applied to mining companies, the mining industry responded by producing codes and standards for implementing EMS and other environmental actions. In these cases, the government had to simply lay the groundwork (Hilson and Nayee, 2002). Voluntary action is encouraged through the creation of industry specific codes and standards that are approved by regulatory agencies. For example, the Minerals Council of Australia developed the Australian Minerals Industry Code for Environmental Management, which was signed by over 43 companies and contains commitments to (Hilson and Nayee, 2002):

- Accept environmental responsibilities for all actions
- Strengthen relationships with surrounding mine communities
- Integrate environmental management into operations
- Minimize the environmental impacts of activities
- Encourage responsible production and use of products
- Continuously improve environmental performance
- Communicate environmental performance

Another example is the Mining Association of Canada's Towards Sustainable Mining initiative, which requires external verification and public disclosure of member performance. The Mining Association of Canada was the first national mining association to require an environmental policy for all members and developed the following guiding principles (Mining Association of Canada, 2014):

- Minimizing impacts to the environment and biodiversity
- Taking responsibility for legacy issues (e.g., abandoned mines)
- Applying best practices and continuous improvements to all aspects of operations
- Encouraging production, re-use, and recycling in an environmentally responsible manner
- Fostering ethical conduct, transparency, and accountability

- Supporting and providing lasting benefits to communities by being responsive to community needs, providing dialogue regarding operations, and enhancing economic, social, educational, environmental, and health care standards
- Respecting human rights, cultures, customs, Aboriginal peoples, and indigenous peoples worldwide

In addition to incorporating industry codes, standards, and regulations into environmental policy, governments can offer incentives. For example, the Queensland Government uses environmental performance criteria to calculate the security deposits required for mining projects. High performers with a consistent track record of environmental protection only pay 25% of estimated rehabilitation liability costs while other companies have to pay up to 100% of the estimated costs (Environment Australia, 2002). Some examples of incentives for adopting an EMS may include (Arimura et al., 2008; Darnall and Pavlichev, 2004):

- Technical and financial assistance for EMS adoption
- Reduced frequency of inspections
- Special recognition or awards
- Expedited permitting
- Preference for public procurement
- Reduced stringency for non-compliance
- Waiver of some environmental regulations

In the United States, industry implementation of EMSs is voluntary. However, the U.S. EPA strongly encourages and provides support for the widespread adoption of EMSs as a flexible method to improve environmental performance and reduce non-compliance throughout the industry (Uchida and Ferraro, 2007). An EMS will not interfere with or replace regulatory programs, but rather, can complement them (U.S. Environmental Protection Agency (U.S. EPA), 2005). In some cases, the U.S. EPA can require the adoption of a compliance focused EMS during enforcement settlements to ensure future compliance to environmental standards (U.S. Environmental Protection Agency (U.S. EPA), 2003).

3.6 IMPLEMENTATION OF EMS

An EMS is implemented using best practices (BPs). The effectiveness of an EMS and BPs depends on existing regulations, administrative and engineered controls, and proper monitoring, assessment, and reviews. Because the adoption of an EMS and BPs is typically not required by regulations, success requires the full support from all levels within the mining company (Environment Australia, 2002).

The EMS implementation process typically involves eight majors steps (National Mining Association, 2012):

1. Project planning
2. Defining policy
3. Identifying environmental aspects and impacts
4. Identifying environmental requirements
5. Assessing environmental impacts and existing process controls
6. Improving controls
7. Monitoring and assessing performance
8. Conducting reviews and developing improvement plans

A typical EMS will take between six to nine months to implement (National Mining Association, 2012), but this may vary depending on the resources committed and the existing environmental programs in place.

Project Planning

Project planning is crucial to the success of any program. During this step, the scope of the EMS is defined, objectives and goals are established, support is obtained from top management, and EMS teams are formed (National Mining Association, 2012). The scope of an EMS should define the operations, sites, and processes that the EMS will cover. The scope may depend on the upstream or downstream processes that will occur and environmental regulations that address these processes.

The objectives of the EMS will vary between companies and depends on the underlying drive for EMS implementation. The primary objective of any EMS should be to reduce environmental risks and improve regulatory compliance; however, other objectives may include improving the company image with stakeholders or obtaining ISO certification (National Mining Association, 2012). Depending on the objectives, the approach taken towards implementing an EMS may differ. For example, if an objective is to receive ISO certification, the EMS may place a much heavier focus on documentation, or if an objective is to improve the company's reputation with stakeholders, a public outreach program may be included as part of the EMS (National Mining Association, 2012).

As with most initiatives, explicit support from top management is necessary to ensure that the EMS has access to the proper resources, focuses on the proper objectives, and is given priority during mining operations.

Without the proper support, an EMS is more likely to fail at meeting the desired objectives.

The final step in project planning is to form an EMS team. An example EMS team may consist of an EMS task force and an EMS coordinator (National Mining Association, 2012). The coordinator should take the lead in managing EMS implementation, leading the task force, and reporting the results to management. The task force should be primarily responsible for implementing the EMS and should consist of members from all major departments within the operation. All members should be given explicit roles and duties for supporting the success of the EMS.

Defining Policy

An environmental policy is a document that provides guidance for an organization's actions and decisions that impact the environment. The policy is usually developed by the EMS coordinator and task force, to be approved by upper management, and acts as a commitment to the EMS, employees, the community, and stakeholders (National Mining Association, 2012). At a minimum, the policy should commit to compliance with regulations, pollution prevention, and continuous improvement (National Mining Association, 2012). The policy should be issued to all employees and be made available to interested parties.

Identifying Environmental Aspects and Impacts

Identifying key activities that may impact the environment or compliance status is vital for developing controls to meet the desired objectives. The EMS task force is typically responsible for this step. The same approach as steps 3 and 4 for the development of an environmental impact assessment (Chapter 2) can be followed to determine activities and operations that have environmental aspects and impacts.

Identifying Environmental Requirements

Once the impacted environmental aspects are identified, it is also important to identify any relevant permits, regulations, ordinances, or laws that are applicable to the operation. Doing so will help ensure regulatory compliance and that EMS objectives are met. This can be done by maintaining and continuously updating a register of legal requirements as well as developing a compliance calendar, which schedules and assigns specific actions to assist with compliance on a day-to-day basis (National Mining Association, 2012).

Assessing Environmental Impacts and Existing Process Controls

Once key activities and legal requirements are identified, environmental aspects can then be thoroughly analyzed to identify aspects of greatest significance in terms of adverse impacts, regulatory compliance, and risk management (National Mining Association, 2012). These aspects will merit robust operational controls and should be considered first.

Environmental aspects should be assessed for relative significance according to three major criteria (National Mining Association, 2012):

- Potential impacts on the environment and human health (both short and long term)
- Applicable regulations and the stringency of regulations
- Potential business impacts (e.g., financial, reputation, etc.)

It is important to note that environmental aspects should be assessed as if there are currently no process controls in place. This ensures an accurate evaluation of the inherit risk posed by each aspect.

Once significant environmental aspects are identified, existing process controls and methods to improve performance can be analyzed. Controls can be classified as either engineered controls (e.g., treatment equipment, containment structures, drainage systems, etc.) or administrative controls (e.g., training, procedures, inspections, maintenance, monitoring, etc.) (National Mining Association, 2012).

Administrative controls should ensure that all employees are properly trained to operate equipment and perform their assigned tasks. Training is important for all employees, as lower level employees who are responsible for day-to-day operations might notice areas for improvement that have gone unnoticed by upper management (Hilson and Nayee, 2002). Standard operating procedures should also be in place for all EMS activities and emergency response plans should be developed to manage unforeseen spills, equipment failure, or accidents. Documentation of training and procedures is also important to analyze past EMS performance and meet regulatory requirements. The efficacy of engineered controls is highly dependent on effective administrative controls. Without proper inspection, maintenance, and operation, engineered controls will eventually fail.

Improving Controls

After assessment is completed, actions need to be taken to improve or establish controls for significant processes to ensure regulatory compliance.

In some cases, there may be no preexisting controls, or existing controls may require additional supporting processes (e.g., document control, reviews, communication systems) (National Mining Association, 2012). For already existing mining operations, areas that need to be improved should be prioritized and dealt with in a systematic manner. Prioritizing areas for improvement helps re-focus mining operations onto sustainable practices such as applying new technologies, rather than trying to treat the system in an end-of-pipe fashion (Newbold, 2006).

Improving controls necessitates the development of an action plan that covers the following elements (National Mining Association, 2012):

- Actions to be taken
  - Designed for integration with current processes when practical
- Responsible party
  - Engagement of personnel involved with the relevant operations
- Required resources
- Timeline

Once a plan is developed and reviewed, approval should be obtained from management to ensure proper resource allocations before implementation.

Monitoring and Assessing Performance

Monitoring and auditing performance is necessary to determine the effectiveness of an EMS at meeting the desired objectives. Monitoring activities should be carried out by the EMS task force on a regular basis according to the compliance calendar. Depending on regulations and permits, specific monitoring activities or sampling methods may need to be followed for certain environmental aspects (National Mining Association, 2012). Monitoring activities may include equipment calibration, routine sampling, laboratory analyses, and visual inspections. Further information on monitoring is provided in Chapter 5.

During monitoring activities, any incidents or non-compliance events should be identified, documented, reported to management, and addressed. What constitutes an "incident" should be defined ahead of time and procedures should already be in place for reporting to management and timeframes for addressing the root causes (National Mining Association, 2012).

Monitoring and tracking performance indicators allows management to understand lapses in performance and take appropriate corrective actions. Performance indicators can be either leading indicators (e.g., employee training, inspections, preventative maintenance, completion of compliance actions, etc.) or lagging indicators (e.g., spills, violations, increased water withdrawal, increased energy usage, increased waste generation, etc.) (National Mining Association, 2012). Leading indicators are directly related to EMS implementation while lagging indicators are usually the result of deficiencies in leading indicators.

Conducting an audit is a formal process for independently and objectively verifying the effectiveness of an EMS. Audits typically involve reviewing work schedules, reviewing documents and records, interviewing staff, and recording on-site observations (National Mining Association, 2012). The results of an audit are reported to site management, facilitating the development and implementation of plans for corrective actions. Formal regulatory audits are usually conducted every two to three years; however, companies may conduct frequent self-audits to ensure compliance. More information on environmental auditing can be found in Chapter 6.

Conducting Reviews and Developing Improvement Plans

The process for conducting reviews should be undertaken by the senior management to ensure continual input and support of the EMS. Management should review the EMS for suitability, adequacy, and effectiveness in the following areas (National Mining Association, 2012):

- Results of audits and compliance with regulations
- Corrective actions and preventative measures
- Environmental performance
- Communications with stakeholders and interested parties
- Adaptability to change

After the review is conducted, new objectives and plans for improvement should be developed. Improvements may be proposed to environmental performance as well as to the EMS itself. New "stretch" goals may be established for long-term improvements or new focus may be placed on the most significant environmental aspects that require improvement (National Mining Association, 2012). During this time, developing action plans and assigning responsibilities for each goal should be done to ensure a clear path forward is established.

REFERENCES

Arimura, T.H., Hibiki, A., Katayama, H., 2008. Is a voluntary approach an effective environmental policy instrument?: a case for environmental management systems. J. Environ. Eco. Manage. 55, 281–295.

Australian Centre for Sustainable Mining Practices, 2011. A Guide to Leading Practice Sustainable Development in Mining. Australia Department of Resources, Energy and Tourism. Retrieved from: http://www.industry.gov.au/resource/Documents/LPSDP/guideLPSD.pdf.

Commission for Environmental Cooperation (CEC), 2005. Successful Practices of Environmental Management Systems in Small and Medium-Size Enterprises: A North American Perspective. Retrieved from: http://www3.cec.org/islandora/en/item/2273-successful-practices-environmental-management-systems-in-small-and-medium-size-en.pdf.

Darnall, N., Pavlichev, A., 2004. Environmental Policy Tools & Firm-Level Management Practices in the United States. Organisation for Economic Co-operation and Development. Retrieved from: http://www.oecd.org/env/consumption-innovation/35590060.pdf.

Environment Australia, 2002. Overview of Best Practice Environmental Management in Mining. Retrieved from: http://commdev.org/userfiles/files/1884_file_Overview_Best_Practice_Environmental_Management_in_Mining20051123111536.pdf.

Fortune Minerals Ltd., 2009. Fortune Minerals & Tahltan Nation Enter into Environmental Assessment Cooperation Agreement for the Mount Klappan Anthracite Coal Project, British Columbia. http://www.fortuneminerals.com/news/press-releases/press-release-details/2009/Fortune-Minerals–Tahltan-Nation-Enter-into-Environmental-Assessment-Cooperation-Agreement-for-the-Mount-Klappan-Anthracite-Coal-Project-British-Columbia/default.aspx.

Hilson, G., Nayee, V., 2002. Environmental management system implementation in the mining industry: a key to achieving cleaner production. Int. J. Mineral Process 64, 19–41.

Holtom, G., 2010. The need for an environmental management system—and what this means for mines. Eng. Min. J., 46–49.

International Atomic Energy Agency, 2010. Best Practice in Environmental Management of Uranium Mining. Retrieved from: http://www-pub.iaea.org/MTCD/publications/PDF/Pub1406_web.pdf.

International Organization for Standardization (ISO), 2013. ISO-1400 Environmental Management. Retrieved from: http://www.iso.org/iso/home/standards/management-standards/iso14000.html.

Ireland Environmental Protection Agency, 2006. Environmental Management Guidelines: Environmental Management in the Extractive Industry (Non-Scheduled Minerals). Retrieved from: http://www.epa.ie/pubs/advice/general/epa_management_extractive_industry.pdf.

Mining Association of Canada, 2014. Towards Sustainable Mining: Progress Report 2014. Retrieved from: http://mining.ca/sites/default/files/documents/TSM_Progress_Report_2014.pdf.

National Mining Association, 2012. Hardrock Mining and Beneficiation Environmental Management System Guide. Retrieved from: http://www.nma.org/pdf/HardrockEMSGuide.pdf.

Natural Resources Canada, 2014. Canada's Mineral Production: Revised Statistics of the Mineral Production of Canada, by Province, 2013. Retrieved from: http://sead.nrcan.gc.ca/prod-prod/2013-eng.aspx.

Newbold, J., 2006. Chile's environmental momentum: ISO 14001 and the large-scale mining industry – Case studies from the state and private sector. J. Clean. Prod. 14, 248–261.

Pacific Institute, 2013. World Water Quality Facts and Statistics. Retrieved from: http://www.pacinst.org/wp-content/uploads/sites/21/2013/02/water_quality_facts_and_stats3.pdf.

Sturm, A., Upasena, S., 1998. ISO 14001 – Implementing an Environmental Management System. Ellipson Management Consultants.

Tole, L., Koop, G., 2013. Estimating the impact on efficiency of the adoption of a voluntary environmental standard: an empirical study of the global copper mining industry. J. Prod. Anal. 39, 35–45.

U.S. Environmental Protection Agency (U.S. EPA), 2003. Guidance on the Use of Environmental Management Systems in Enforcement Settlements as Injunctive Relief and Supplemental Environmental Projects. Retrieved from: https://www.fedcenter.gov/_kd/Items/actions.cfm?action=Show&item_id=2584&destination=ShowItem.

U.S. Environmental Protection Agency (U.S. EPA), 2005. United States Environmental Protection Agency Position Statement on Environmental Management Systems (EMSs). Retrieved from: https://www.fedcenter.gov/_kd/Items/actions.cfm?action=Show&item_id=3795&destination=ShowItem.

U.S. Environmental Protection Agency (U.S. EPA), 2013. Environmental Management Systems (EMS). http://www.epa.gov/ems/#iso14001.

Uchida, T., Ferraro, P.J., 2007. Voluntary development of environmental management systems: motivations and regulatory implications. J. Regul. Econ. 32, 37–65.

United Nations, 2002. Berlin II Guidelines for Mining and Sustainable Development. Retrieved from: http://commdev.org/userfiles/files/903_file_Berlin_II_Guidelines.pdf.

CHAPTER 4

Environmental Impacts of Mining

4.1 INTRODUCTION

The environmental characteristics, or attributes, defined and described in this chapter represent a selected set of elements designed to be used in the environmental assessment process for mining and mineral processing. These attributes may be applicable for certain mining situations, but may not be all inclusive. Environmental attributes should be adapted to each specific mining situation. Five primary environmental areas of concern were selected relating to mining and mineral processing, with a sixth, economics, selected as a discussion topic as mining is closely tied with the economy of a country. They are:

- Air quality
- Water quality/quantity
- Acid mine drainage
- Land impacts
- Ecological impacts
- Economic impacts

In reality, all six areas are closely tied to one another and it can become difficult to separate one from the other.

4.2 AIR QUALITY IMPACTS FROM MINING

Air emissions from mining can affect not only the surrounding areas, but can affect regional and global air quality. Air quality can generally be defined as the measure of ambient atmospheric pollution relative to the potential to inflict environmental harm or to adversely affect human health. Air quality is highly vulnerable because unlike water or other wastes, air cannot, in practice, be reprocessed at some central location and subsequently distributed for reuse (Jain et al., 2012). Because of this vulnerability, air pollution and particulate matter (PM) emissions should be controlled and decreased at the source to reduce negative impacts on nearby communities, local flora and fauna, and the global environment. In order to reduce air quality impacts, it is important to understand how mining

Environmental Impact of Mining and Mineral Processing
ISBN 978-0-12-804040-9
http://dx.doi.org/10.1016/B978-0-12-804040-9.00004-8

processes contribute to air pollution and potential impacts such emissions may have on the environment. This section will cover the direct impacts of mining and mineral processing on air quality.

Air Quality Legislation

In order to fully understand the implications of air pollution from the mining and mineral processing industry, it is necessary to have a brief background in air quality standards and legislation. In the United States, air quality is regulated under the Clean Air Act (42 U.S.C. 7401 et seq.). Other industrialized countries have similar standards that should be followed accordingly. The primary goal of the Clean Air Act is to "protect and enhance the quality of the Nation's air resources so as to promote the public health and welfare and the productive capacity of its population." In the United States, the Clean Air Act was originally passed in 1970 and has since been amended to cover issues such as acid rain and other toxic pollutants. Under the authority of the Clean Air Act, the U.S. EPA established National Ambient Air Quality Standards for outdoor air that provides limits for sulfur dioxide, ozone, nitrogen oxide, particulate matter, carbon monoxide, and lead (Table 4.1). Primary and secondary standards are provided for public health protection (e.g., children, elderly, sensitive groups), and public welfare protection (e.g., visibility, transportation, effects on vegetation, animals, soils, or water), respectively. The Clean Air Act also designates areas that do not meet the standards as nonattainment areas and requires the submittal of plans for reaching attainment. These plans must include information on all pollutants, permits, control measures, and contingency measures.

The U.S. EPA has estimated that the Clean Air Act has prevented 160,000 cases of adult mortality and 1.7 million cases of asthma in the year 2010, and projects these numbers to reach 230,000 and 2.4 million for adult mortality and asthma, respectively, for 2020 (U.S. Environmental Protection Agency (U.S. EPA), 2011). However, as with any legislation, compliance places an economic burden on pollution emitters. By the year 2020, is it estimated that the direct cost of Clean Air Act compliance will reach $65 billion nationwide; however, the estimated direct benefits of the act will reach $2 trillion (U.S. Environmental Protection Agency (U.S. EPA), 2011). Similar studies have been conducted, and are being conducted, in other industrialized countries with comparable results.

The European Union manages air quality through air quality limit values, originally established in 1980 under Directive 80/779/EEC.

Table 4.1 Primary and secondary standards for criteria pollutants as set by the U.S. EPA in the National Ambient Air Quality Standards (40 CFR part 50) (U.S. Environmental Protection Agency (U.S. EPA), 2014c)

Pollutant		Primary/Secondary	Time	Limit	Form
Carbon monoxide		Primary	8 h	9 ppm	Not to be exceeded more than once per year
			1 h	35 ppm	
Lead		Primary and secondary	Rolling 3 month average	0.15 µg/m^3	Not to be exceeded
Nitrogen dioxide		Primary	1 h	100 ppb	98th percentile of 1 h daily maximum concentration, averaged over 3-year period
		Primary and secondary	Annual	53 ppb	Annual mean
Ozone		Primary and secondary	8 h	0.075 ppm	Annual fourth highest daily maximum 8-h concentration, averaged over 3-year period
Particulate matter	PM$_{2.5}$	Primary	Annual	12 µg/m^3	Annual mean, averaged over 3-year period
		Secondary	Annual	15 µg/m^3	Annual mean, averaged over 3-year period
		Primary and secondary	24 h	35 µg/m^3	98th percentile, averaged over 3-year period
	PM$_{10}$	Primary and secondary	24 h	150 µg/m^3	Not to exceed, on average, more than once per year, over 3-year period
Sulfur dioxide		Primary	1 h	75 ppb	99th percentile of 1 h daily maximum concentration, averaged over 3 years
		Secondary	3 h	0.5 ppm	Not to be exceeded more than once per year

Additional Directives over the years expanded on air quality limit values for addition constituents. The relatively late timeframe of developing region wide air quality standards meant that many member states had already implemented their own legislation (Milieu Ltd et al., 2004). Thus, the 1996 Air Quality Framework Directive was established to standardize the management of air quality throughout the European Union. The current air quality limit values are provided in Table 4.2.

Other countries have implemented similar legislation regarding air quality. Both the United Kingdom and Canada have respective Clean Air Acts. China regulates air quality through the Law of the People's Republic of China on the Prevention and Control of Atmospheric Pollution and has recently adopted an Air Pollution Prevention and Control Action Plan to control particulate matter and ozone emissions.

Internationally, there are other agreements concerning general air quality and atmospheric protection such as the 1979 Geneva Convention on Long-rang Transboundary Air Pollution and its protocols concerning acidification, heavy metals, persistent organic pollutants, sulfur, volatile organic compounds (VOCs), and NO_x. The Convention specifies emission reduction obligations for the respective countries that participate, rather than a global standard for specific pollutants. Additionally, the World Health Organization has issued air quality guidelines for six key air pollutants (Table 4.3). Other international agreements target specific areas of

Table 4.2 European Union air quality guidelines (European Union, 2015)

Pollutant	Time	Limit
$PM_{2.5}$	Annual mean	$25~\mu g/m^3$
PM_{10}	Annual mean	$40~\mu g/m^3$
	24 h mean	$50~\mu g/m^3$
Ozone	8 h mean	$120~\mu g/m^3$
Nitrogen dioxide	Annual mean	$40~\mu g/m^3$
	1 h mean	$200~\mu g/m^3$
Sulfur dioxide	24 h mean	$125~\mu g/m^3$
	1 h mean	$350~\mu g/m^3$
Carbon monoxide	8 h mean	$10~mg/m^3$
Lead	Annual mean	$0.5~\mu g/m^3$
Benzene	Annual mean	$5~\mu g/m^3$
Arsenic	Annual mean	$6~ng/m^3$
Cadmium	Annual mean	$5~ng/m^3$
Nickel	Annual mean	$20~ng/m^3$
Polycyclic aromatic hydrocarbons	Annual mean	$1~ng/m^3$ (as concentration of Benzo(a)pyrene)

Table 4.3 World Health Organization air quality guidelines
(World Health Organization, 2010)

Pollutant	Time	Limit
$PM_{2.5}$	Annual mean	10 μg/m^3
	24 h mean	25 μg/m^3
PM_{10}	Annual mean	20 μg/m^3
	24 h mean	50 μg/m^3
Ozone	8 h mean	100 μg/m^3
Nitrogen dioxide	Annual mean	40 μg/m^3
	1 h mean	200 μg/m^3
Sulfur dioxide	24 h mean	20 μg/m^3
	10 min mean	500 μg/m^3
Carbon monoxide	30 min mean	60 mg/m^3
	1 h mean	30 mg/m^3
	8 h mean	10 mg/m^3

concern such as acid rain (Air Quality Agreement between the U.S. and Canada), ozone (Vienna Convention for the Protection of the Ozone Layer), and greenhouse gases (Framework Convention on Climate Change and the Kyoto Protocol).

Air Quality Impacts from Mining and Mineral Processing

Air pollutants can be roughly categorized as either gaseous or particulates. Gaseous emissions of concern from mining and mineral processing activities include sulfur oxides, nitrogen oxides, carbon oxides, photochemical oxidants, VOCs, hydrocarbons, and methane. Gaseous emissions typically originate from mining equipment and processes, including diesel motors and blasting. However, some gaseous pollutants, such as methane, originate from the mineral deposits. Particle emissions contribute to the majority of air quality problems at a mining site and primarily result from land clearing and removal, excavation, ore crushing, loading, and vehicle transport. Air emission sources are classified as mobile, stationary, or fugitive emissions sources. Mobile sources include vehicles, trucks, and other excavation equipment, where emission reduction (if any) is incorporated into the vehicle. Stationary sources include large operations such as drying, roasting, and smelting, where gaseous emissions are typically filtered or treated before venting. Fugitive emissions are emissions from various activities, such as materials handling and storage, fugitive dust, and blasting that cannot be reasonably passed through a vent, stack, or chimney system to reduce emissions (Environmental Law Alliance Worldwide, 2010).

Table 4.4 provides a generalized description of typical mining activities that are responsible for gaseous, particulate, and odorous emissions. Because every mine site will vary in geology and activity, Table 4.4 will vary from mine to mine. Table 4.5 summarizes major air pollutants of concern and typical sources from mining operations.

Methane Emissions

Methane, CH_4, is a colorless, odorless gas that is naturally present in coal seams and is the primary component of natural gas. During the formation of coal, methane gas is produced and is stored within the internal surfaces of the coal (U.S. Geological Survey (USGS), 2000). Methane occurs with most coals and can be released as a by-product at great volumes during coal mining operations, or due to natural erosion. The release of coal-mine methane (CMM) poses an immediate explosive and respiratory hazard to underground workers. Methane is highly explosive, with lower and upper explosive limits of 5% and 15% concentration by volume, respectively, and also displaces oxygen in the atmosphere, acting as an asphyxiant (U.S. Environmental Protection Agency (U.S. EPA), 2009b).

In order to protect mine workers from the hazards of CMM, underground mines are required to have ventilation systems installed to provide adequate air flow in order to dilute and remove methane gas. Methane gas from such systems is in very low concentrations (typically <1%) and is often vented to the atmosphere (U.S. Environmental Protection Agency (U.S. EPA), 2009b).

One method to avoid the release of CMM during mining is to pre-drain methane, called degasification, or to post-drain methane, called "gob" (Banks, 2012). Both systems require the drilling of boreholes to allow CMM to naturally vent out of the coal seam. This methane is typically of high enough quality to be sold or used on-site for energy generation, but can also be vented.

Once in the atmosphere, methane acts as a greenhouse gas and is more efficient at trapping radiation than carbon dioxide. Although methane has an atmospheric life of 12 years, its overall impact on climate change is more than 20 times that of carbon dioxide. Methane is one of six greenhouse gases, along with carbon dioxide, nitrous oxide, hydrofluorocarbons, perfluorocarbons, and sulfur hexafluoride, addressed by the United Nations Kyoto Protocol for emission reduction.

Table 4.4 Air quality impacts from open and underground mining

Activity/Source	Coal mining		Metalliferous mining	
	Surface cut	Underground	Surface cut	Underground
Earthmoving associated with construction and development of surface facilities	P	P	P	P
Shaft/drift access and ventilation development		P		P
Removing vegetation and topsoil for mine preparation	P		P	
Drilling and blasting*	P, G, O		P, G, O	
Removing and placing overburden	P		P	
Extracting, transporting and dumping coal or ore	P	P	P	P
Crushing coal, ore, and other materials	P	P	P	P
Screening	P	P	P	P
Washery options				
Beneficiation of material**			P, G, O	P, G, O
General materials handling	P	P, G	P	P, G
Transporting and placing washery rejects	P	P		
Workshop and/or power plant operations***	P, G, O	P, G, O	P, G, O	P, G, O
Rehabilitation	P	P	P	P
Wind erosion from open pit, stockpiles, and exposed areas (including tailings)	P	P	P	P
Rail transport	P	P	P	P
Ship loading	P	P	P	P

This table provides a general guideline and should be adapted based on site-specific properties and practices.

P = Dust or primary particle emissions occur.

G = Significant gaseous emissions occur.

O = Significant odorous emissions are possible.

* Gaseous pollutant of most concern is nitrogen dioxide.

** Pollutant typically of most concern is sulfur dioxide. For gold ore processing, hydrogen cyanide can be emitted.

*** Gaseous emissions can include sulfur dioxide and various other volatile organic compounds.

Adapted from Leading Practice Sustainable Development Program for the Mining Industry (2009).

Table 4.5 Pollutants affecting air quality

Air pollutant	Effect on health or environment	Source
Carbon monoxide	Combines with hemoglobin in blood, displacing oxygen that hemoglobin normally transports, reducing oxygen carrying capacity of the circulatory system. Results in reduced reaction time, increased burden on pulmonary system in cardiac patients, and death with prolong exposure.	Mining equipment
Carbon dioxide	In the atmosphere, increases the trapping of long-wave radiation, causing the atmosphere to become warmer. Many aspects of climate change follow from this increased temperature.	Mining equipment
Photochemical oxidants	Nitrogen oxides react with hydrocarbons in the presence of sunlight to form photochemical oxidants; causes eye, ear, and noise irritation and adversely affects plant life.	Atmospheric reactions
Hydrocarbons	Combines with oxygen and nitrogen oxides to form photochemical oxidants.	Mining equipment
VOCs	Can cause short-term or long-term health issues. Can combine with other chemicals in the atmosphere.	Mining equipment, fuels, and solvents
Nitrogen oxides	Form photochemical smog and are associated with a variety of respiratory diseases.	Mining equipment
Sulfur dioxide	Associated with respiratory diseases and can form compounds resulting in corrosion and plant damage.	Mining equipment
Methane	Highly flammable, becomes explosive at certain mixtures with air. Can cause asphyxia as methane displaces oxygen.	Naturally occurring with coal
Hazardous toxicants	Can adversely affect health at very low exposure levels.	Lead, mercury, and other heavy metals, crystalline silica, asbestos, radio nuclides.
Particulate matter (PM)	PM can become trapped in lungs, creating breathing problems. Can cause maintenance issues and be an annoyance to local communities.	Many mining activities

Adapted from Jain et al. (2012), Nevada Mining Association (2010), and Leading Practice Sustainable Development Program for the Mining Industry (2009).

Methane Emissions from Coal Mines

Methane released from coal mining activities comes from five major sources (Figure 4.1) (U.S. Environmental Protection Agency (U.S. EPA), 2014b):

- Surface mines where seams are directly linked to the atmosphere
- Degasification systems from underground mines via vertical or horizontal wells
- Ventilation air from underground mines
- Fugitive emissions from coal during storage, transportation, and other post–mining activities
- Abandoned or closed mines that leak methane through vent holes or fissures

The quantity of methane released is depended on the methane content of the coal and the coal depth. Coals with higher carbon content (rank) tend to have higher concentrations of methane and deep coal seams are usually under greater pressure, preventing the natural migration of methane out of the coal resulting in greater releases during mining (U.S. Environmental Protection Agency (U.S. EPA), 2012b). Thus, surface mining typically releases less methane than underground mines due to the lower rank and depth of the coal.

Ventilation emissions from underground coal mines represent the largest source of coal mine methane emissions, both in the United States and

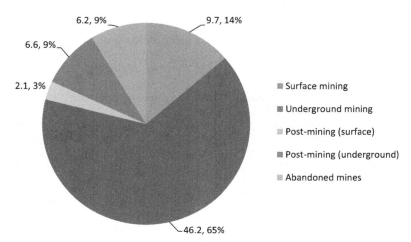

Figure 4.1 Breakdown of methane emissions from coal mining activities in the U.S. in 2013 (U.S. Environmental Protection Agency (U.S. EPA), 2015b). Units are measured in $MtCO_2Eq$ (million metric tons of carbon dioxide equivalent). Underground mining category includes both ventilation and degasification emissions.

around the world. Methane release from abandoned mines, although significant, vary greatly and are dependent on the time since abandonment, whether the mine is flooded or not, gas content of the coal seam, and the presence of mine seals and vent holes (U.S. Environmental Protection Agency (U.S. EPA), 2015b).

Methane releases from mining activities can be estimated using methane emission factors for different types of operations, as provided by the Intergovernmental Panel on Climate Change (IPCC) (2006) (Table 4.6). The emission factors provided are general guidelines; country or mine specific data should be used if available.

Methane emissions from surface and underground coal mines, as well as post mining emissions can be calculated using the following formula (Intergovernmental Panel on Climate Change, 2006):

$$ME = (MEF)(P)(CF)$$

where:

ME = methane emissions, Gg year^{-1}

MEF = methane emission factor, m^3 ton^{-1}

P = coal production, ton year^{-1}

CF = conversion factor for converting volume of CH_4 to mass at 20°C and 1 atm; 0.67×10^{-6} Gg m^{-3}

To determine the total methane emissions from surface mining activities, both the mining and post-mining emissions must be taken into account. Total methane emissions can be calculated by summing the respective methane emissions calculated using below equation:

$$TME = ME_{Surface} + ME_{Post-mining(surface)}$$

Table 4.6 Methane emission factors for various coal mining operations (Intergovernmental Panel on Climate Change, 2006)

	Underground mining	Post mining (underground)	Surface mining	Post mining (surface)
Low emission factor	10	0.9	0.3	0
Average emission factor	18	2.5	1.2	0.1
High emission factor	25	4.0	2.0	0.2

All emission factors are given in units of m^3 ton^{-1}.

where:

TME = total methane emissions, Gg year^{-1}

$ME_{Surface}$ = methane emissions from surface mining, Gg year^{-1}

$ME_{Post-mining(surface)}$ = post-surface mining methane emissions, Gg year^{-1}

In underground mining, methane emissions can sometimes be recovered and utilized for energy production, or flared. In cases where a portion of the methane is recovered and utilized for energy, or flared, the equation becomes:

$$TME = ME_{Underground} + ME_{Post-mining(Underground)} - VM_{recovered} + ME_{flaring}$$

where:

TME = total methane emissions, Gg year^{-1}

$ME_{Underground}$ = methane emissions from underground mining, Gg year^{-1}

$ME_{Post-mining(Underground)}$ = post-underground mining methane emissions, Gg year^{-1}

$VM_{recovered}$ = volume of methane recovered or flared, Gg year^{-1}

$ME_{flaring}$ = methane emissions from flaring, Gg year^{-1}

In cases where a portion of the methane is flared, emissions of unburned methane must be taken into account and can be calculated using the following formula:

$$ME_{flaring} = (0.02)(V_{flared})(CF)$$

where:

V_{flared} = volume of methane flared, Gg year^{-1}

Methane emissions from abandoned wells can be calculated using the following formula:

$$ME_{abandoned} = (N)(G)(ER_{avg})(MEF)(CF)$$

where:

$ME_{abandoned}$ = methane emissions abandoned mines, Gg year^{-1}

N = number of abandoned coal mines that are unflooded, unitless

G = the fraction of gassy mines, unitless

ER_{avg} = the average emission rate, m^3 year^{-1}

In flooded mines, the water will prevent future methane release and emissions can be considered negligible for this calculation. If no data is available on the state of abandoned mines, the safe estimation is to assume all mines are unflooded. The fraction of gassy coal mines is dependent on the time period in which the mine was abandoned according to Table 4.7. Emission factors for abandoned gassy coal mines

Table 4.7 Percentage of abandoned gassy coal mines in relation to time (Intergovernmental Panel on Climate Change, 2006)

Time interval	Low	High
1900–1925	0%	10%
1926–1950	3%	50%
1950–1976	5%	75%
1976–2000	8%	100%
2001–Present	9%	100%

depend on both the inventory year and the time interval of closure, and can be found in Appendix A, Table A4.

Further guidelines and alternate methods for calculating methane emissions from coal mining operations are provided in the IPCC Guidelines for National Greenhouse Gas Inventories, but are dependent on the amount of data available (e.g., historical emissions, mine-specific or basin-specific emissions factors, type of coal, etc.).

In 2010, coal mine emissions represented 8%, or 584 $MtCO_2e$ (million metric tons of carbon dioxide equivalent), of total global emissions of methane (Global Methane Initiative, 2011). China accounted for a large percentage of coal mining methane emissions with 299.5 $MtCO_2e$ in 2010, followed by the United States and Russia with 59.0 and 55.2 $MtCO_2e$, respectively (Table 4.8).

Table 4.8 Estimated coal methane emissions and coal production for select countries in 2010 (Global Methane Initiative, 2011; U.S. Energy Information Administration, 2013)

Country	Coal methane emissions ($MtCO_2e$)	Coal production (Mt)	Methane emission per coal production ($MtCO_2e/Mt$)
China	299.5	3230.16	0.0927
United States	59.0	983.72	0.0600
Russia	55.2	321.70	0.1716
Ukraine	27.4	54.95	0.4986
Australia	26.8	424.40	0.0631
India	26.5	562.31	0.0471
Kazakhstan	13.5	110.93	0.1217
Poland	8.3	132.68	0.0626
Colombia	7.4	74	0.0995
Vietnam	6	44.83	0.1338

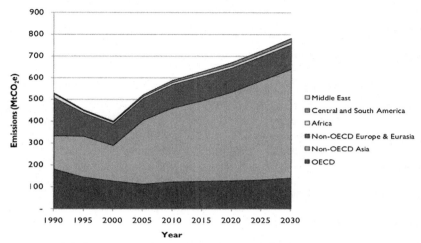

Figure 4.2 Global projections for methane emissions from coal mining activities (U.S. Environmental Protection Agency (U.S. EPA), 2012b).

The U.S. EPA predicts that methane emissions from coal mining will increase by approximately one third to 784.3 $MtCO_2e$ by 2030 (Figure 4.2) (U.S. Environmental Protection Agency (U.S. EPA), 2012b). Much of the increase in global emissions is expected to come from non-OECD Asian countries, particularly China due to its rapid economic growth, abundant coal resources, and increasing demand for energy (U.S. Environmental Protection Agency (U.S. EPA), 2012b).

Methane Emissions from Non-coal Mines

Although the majority of methane emissions from mining and mineral processing comes from coal mines, methane is also sometimes associated with salt, potash, gold, lead, diamond, trona, and other base metals mining (U.S. Environmental Protection Agency (U.S. EPA), 2013). In these cases, mineral deposits are located near methane containing deposits, such as coal or carbonaceous shales, where the methane originates (U.S. Environmental Protection Agency (U.S. EPA), 2013). Methane can migrate in gaseous form through joints, dykes, or faults, or it can be dissolved in water under pressure and transported into underground voids (Head and Kissell, 2006). Methane emissions rates in non-coal mines are inconsistent and can vary between regions, making predicting, measuring, and managing these methane emissions difficult. Examples of non-coal mining operations with known methane emissions can be found in Table 4.9.

Table 4.9 Non-coal mining operations with methane emissions according to country or region (Head and Kissell, 2006)

Country or region	Non-coal mining operations with reported methane emissions
Scandinavia	Iron
Eastern Europe	Igneous and metamorphic rock
Romania	Mica schist
Hungary	Copper
United Kingdom	Granite, iron
Canada	Canadian shield mines (nickel, gold, silver, copper)
United States	Oil shales, salt, trona, potash, limestone, copper, uranium
Australia	Copper
South Africa	Gold, platinum, kimberlite pipes

Sulfur Dioxide Emissions

Sulfur dioxide (SO_2) is a colorless, reactive gas with a strong odor. Sulfur dioxide comes from a variety of natural and anthropogenic sources. The primary anthropogenic sources of sulfur dioxide emissions are the burning of high-sulfur coals and heating oils in power plants, followed by industrial boilers and metal smelting. Natural causes contribute anywhere from 35–65% of total sulfur dioxide emissions annually and include sources such as volcanoes (World Bank Group, 1998).

When released into the atmosphere, sulfur dioxide can react to form acid rain according to the following reactions:

$$SO_2 + OH\cdot \rightarrow HOSO_2\cdot$$

$$HOSO_2\cdot + O_2 \rightarrow HO_2\cdot + SO_3$$

$$SO_3 \text{ (g)} + H_2O \text{ (l)} \rightarrow H_2SO_4 \text{ (aq)}$$

or

$$SO_2 \text{ (g)} + H_2O \leftrightarrow SO_2\cdot H_2O$$

$$SO_2\cdot H_2O \leftrightarrow H^+ + HSO_3^-$$

$$HSO_3^- \leftrightarrow H^+ + SO_3^{2-}$$

Sulfur dioxide has a direct impact on human health and is responsible for a variety of respiratory problems (Table 4.10). Like most air pollutants, sulfur dioxide poses a greater threat to sensitive groups such as the elderly,

Table 4.10 Health and environmental effects of sulfur dioxide and its reaction products (Center for Research Information, 2004; U.S. Environmental Protection Agency (U.S. EPA), 2012a; World Bank Group, 1998)

Pollutant	Type of impact	Effects
Acid rain	Environmental	Acidification of lakes and streams, release of aluminum from soils, reduced fish growth or death, reduced plant growth, damage to leaves, dissolves and washes away nutrients and minerals
	Materials	Damage to stone, concrete, paint, buildings
Sulfur dioxide	Visibility	Haze, reduced visibility
	Plants	Reduced plant growth, damage to leaves, premature death
	Human health	Irritation of skin, eyes, lungs, throat, and nose, bronchoconstriction, coughing, shortness of breath, increased asthma symptoms, inflammation of respiratory system, unpleasant odor

asthmatics, and young children. By contributing to acid rain, sulfur dioxide can have significant impacts on plants, surface waters, and buildings. However, acid rain is not considered to have direct human health impacts (U.S. Environmental Protection Agency (U.S. EPA), 2012a). Sulfur dioxide and other sulfur oxides can also react in the atmosphere to form particulate matter and ground level ozone further contributing to negative health effects (See sections on VOCs and PM for more information on ozone and PM, respectively).

In the United States, sulfur dioxide emissions totaled 5.86 million metric tons in 2011, of which fuel combustion accounted for 84% (4,920,849 million metric tons) of the total, followed by industrial processes at 10.3% (605,228 million metric tons) (U.S. Environmental Protection Agency (U.S. EPA), 2015a). Of the industrial sources, non-ferrous metal and ferrous metal processing accounted for approximately 16% and 4% of emissions respectively (Figure 4.3). General mining processes accounted for less than 1%.

Since 2006, global sulfur dioxide emissions have been decreasing, primarily due to decreased emissions from the U.S. and Europe and the implementation of stricter emission limits from power plants in China (Klimont et al., 2013). However, some regions, such as India, Africa, and the Middle East continue to have increased sulfur dioxide emissions.

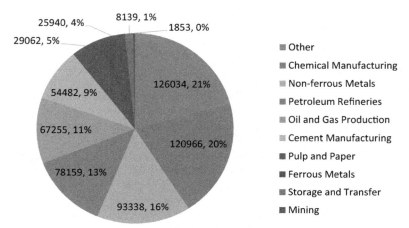

Figure 4.3 Emissions of sulfur dioxide in the United States from industrial processes for 2011 (U.S. Environmental Protection Agency (U.S. EPA), 2015a). Units are in metric tons.

During mining and mineral processing, sources of sulfur dioxide emissions include vehicle exhaust from fuel combustion and metal smelting. Many types of vehicles including bulldozers, excavators, drills, graders, trucks, and front end loaders are indispensable to mining operations and thus, vehicle emissions are unavoidable. These vehicles typically use diesel fuel, the combustion of which produces SO_x, as well as other air pollutant such as CO, NO_x, VOCs, and PM. The other air pollutants are discussed in later sections.

Many metals are naturally found as sulfide ores, including iron, copper, zinc, lead, mercury, molybdenum, arsenic, antimony, and sometimes gold. As a result, sulfur oxides are produced during metal processing, typically during the roasting or sintering stages. For example, during the roasting of zinc ore (zinc sulfide), the ore is heated in the presence of oxygen according to the following reactions, forming zinc oxide and sulfur oxides (U.S. Environmental Protection Agency (U.S. EPA), 1995e):

$$2 \text{ ZnS} + 3 \text{ O}_2 \rightarrow 2 \text{ ZnO} + \text{SO}_2$$

$$2 \text{ SO}_2 + \text{O}_2 \rightarrow 2 \text{ SO}_3$$

Zinc roasting emits approximately 93–97% of the sulfur content in the feed as sulfur oxides and accounts for nearly 90% of potential sulfur oxide emissions in zinc processing (U.S. Environmental Protection Agency (U.S. EPA), 1995e). The concentration of sulfur dioxide in emissions from common smelting operations is provided in Table 4.11.

Table 4.11 Concentration of sulfur dioxide in emission from copper, lead, and zinc smelting (U.S. Environmental Protection Agency (U.S. EPA), 1995b,c,e)

Process	Concentration of SO_2 (%)
Primary copper smelting processes	
Multiple hearth roaster	1.5—3
Fluidized bed roaster	10—12
Reverberatory furnace	0.5—1.5
Electric arc furnace	4—8
Flash smelting furnace	10—70
Continuous smelting furnace	5—15
Pierce-Smith converter	4—7
Hoboken converter	8
Single contact H_2SO_4 plant	0.2—0.26
Double contact H_2SO_4 plant	0.05
Lead smelting processes	
Sintering (Front end gas)	2.5—4
Zinc smelting processes	
Multiple hearth roasters	4.5—6.5
Suspension roasters	10—13
Fluidized bed roasters	7—12

Sulfur dioxide emissions from vehicles and general mining and mineral processing can be calculated using standard emission factors.

$$SDE_{process} = \left(SDEF_{process} \right)\left(Unit_{process} \right)$$

where:

$SDE_{process}$ = sulfur dioxide emissions for a given process; lbs or kg

$SDEF_{process}$ = sulfur dioxide emission factor for given process; lbs $unit^{-1}$ or kg $unit^{-1}$

$Unit_{process}$ = units for the given process; tons processed, tons produced, ton charged, tons transferred, miles traveled, etc.

Sulfur dioxide emission factors for select mining processes can be found in Appendix A, Tables A1 and A2. Emission factors in Appendix A are for uncontrolled emissions. If emission control equipment is in place, they must be taken into account when calculating emissions.

Volatile Organic Compound Emissions

Volatile organic compounds (VOCs) represent a diverse class of organic compounds with high vapor pressures at room temperature. These compounds can be man-made or naturally occurring (biogenic). Volatile organic compounds are widely used in consumer products (e.g., personal care products, paint, degreasers, refrigerants, aerosols, furniture, etc.) and in industry as solvents, for chemical manufacturing, and are the byproduct of the combustion of gasoline, diesel fuel, and the burning of coal. Examples of VOCs include benzene, formaldehyde, toluene, and xylene. Volatile organic compounds are of concern to both indoor and outdoor air quality.

Indoors, the primary concern is inhalation exposure to VOCs due to the resulting negative health effects. These health effects can vary greatly, depending on the specific compound, length of exposure, and individual susceptibility. Some compounds have no known health effects, while others can be highly toxic or carcinogenic.

Outdoors, VOC emissions are of concern primarily due to their ability to create photochemical smog. When released into the atmosphere, VOCs can react with nitrogen oxides (NO_x), hydroxyl radicals ($OH\cdot$), and ozone (O_3) (Arey and Atkinson, 2006). In the presence of sunlight, VOCs and NO_x can form ozone, a regulated air pollutant, at ground levels. Other atmospheric transformations may lead to the formation of secondary organic aerosols, particulate matter, and toxic air contaminants such as peroxyacetyl nitrates (Arey and Atkinson, 2006). Furthermore, the reaction of VOCs with hydroxyl radicals in the atmosphere reduces the availability of hydroxyl radicals, leading to the accumulation greenhouse gasses such as methane (European Science Foundation, 2005; Sahu, 2012). Volatile organic compounds interact with hydroxyl radicals and nitrogen oxide species according to the following reactions (Sahu, 2012):

$$VOCs + OH\cdot + O_2 \rightarrow RO_2 + H_2O$$
$$RO_2 + NO + O_2 \rightarrow NO_2 + HO_2 + CARB$$
$$HO_2 + NO \rightarrow NO_2 + OH\cdot$$
$$2(NO_2 + hv + O_2 \rightarrow NO + O_3)$$

$$\overline{Net : VOCs + 4\,O_2 \rightarrow 2\,O_3 + CARB + H_2O}$$

where CARB is carbonyl compounds and hv is solar radiation.

The U.S. EPA defines VOCs for outdoor air pollution as "any compound of carbon, excluding carbon monoxide, carbon dioxide, carbonic acid, metallic carbides or carbonates, and ammonium carbonate, which participates in atmospheric photochemical reactions" (40 CFR 51.100). Other compounds not considered VOCs due to their lack of photochemical reactivity include methane, ethane, acetone, methylene chloride, some Freon compounds, perchloroethylene, and other halogenated hydrocarbons (40 CFR 51.100).

Volatile organic compounds are a primary contributor to photochemical smog. Photochemical smog consists of a mixture of ozone, peroxyacyl nitrates, organic nitrates, oxidized hydrocarbons, VOCs, and other photochemical aerosols and particulates. Photochemical smog, in addition to creating a brown haze over cities, can result in negative health effects, many of which are worse for vulnerable groups such as asthmatics and the elderly (Table 4.12) (Environment Protection Authority, 2004). Furthermore, some components such as ozone can negatively impact plant growth and can damage various man-made materials (e.g., rubber, textiles, paint) (Environment Protection Authority, 2004).

Global emissions of VOCs are estimated at 1210–1290 TgC (teragrams of carbon) annually (Sahu, 2012). The emission of biogenic VOCs, primarily emitted from plants, is estimated at 1150 TgC annually on a global scale, and exceed anthropogenic sources by an order of magnitude (European Science Foundation, 2005; Sahu, 2012). Anthropogenic emissions of VOCs are estimated to be 60–140 TgC annually (Sahu, 2012). Worldwide, the majority of anthropogenic VOC emissions come from solvent use, industrial processes, road transportation, and commercial institutions and households (Evuti, 2013) (Figure 4.4).

Table 4.12 Health effects of exposure to constituents of photochemical smog (Environment Protection Authority, 2004; Gray and Finster, 1999; U.S. National Library of Medicine, 2014)

Pollutant	Population	Health effects	
Nitrogen oxides	Humans	See Table 4.13	
Ozone	Humans	Coughing, wheezing, eye irritation, reduced lung function, inflame lung tissue, acute respiratory problems, impaired immune response, aggravate existing conditions	
	Plants	Premature mortality, interferes with photosynthesis, reduced growth, leaf damage, reduced agricultural yields	
VOCs	Humans	Long-term	Damage to liver, kidneys, and nervous system, cancer
		Short-term	Eye irritation, respiratory tract irritation, headaches, fatigue, dizziness, skin reactions, nausea, visual disorders, memory impairment
Peroxyacetyl nitrate	Humans	Eye irritation, respiratory problems	

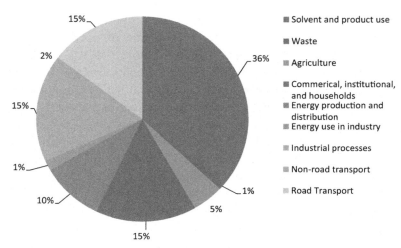

Figure 4.4 Relative contributions to global emissions of non-methane VOCs by sector (Evuti, 2013).

Emissions of VOCs can come from a variety of sources during mining and mineral processing. The most common sources of VOC emissions include:
- Vehicle exhaust and equipment leaks
- Solvents for cleaning and degreasing
- Fuel storage and combustion

Volatile organic compound emissions for vehicle exhaust and typical mining processes can be estimated using the following equation:

$$VOCE_{process} = \left(EF_{process}\right)\left(Unit_{process}\right)$$

where:

$VOCE_{process}$ = volatile organic compound emissions for a select process; lbs or kg

$EF_{process}$ = emission factor for specific process; lbs unit^{-1}, kg unit^{-1}, Mg unit^{-1}

$Unit_{process}$ = units for given process; tons processed, tons produced, ton charged, tons transferred, miles traveled, kg solvent used, etc.

Volatile organic compound emission factors for select mining processes can be found in Appendix A, Table A1 and emission factors of various mining equipment exhaust can be found in Appendix A, Table A2. Emission factors in Appendix A are for uncontrolled emissions. If emission control equipment is in place, they must be taken into account when calculating emissions.

Emissions of VOCs from cleaning and degreasing operations are associated with maintenance activities. Although control equipment can increase the recovery of solvents, estimates for overall process emissions typically assume that 100% of the volatile solvents consumed will eventually be released to the atmosphere (National Pollutant Inventory, 1999). If consumption information is unavailable, emission rates for degreasing operations can be estimated on a per unit basis, as provided in Appendix A, Table A3.

Emissions of VOCs from fuel storage and transportation are primarily the result of evaporative losses from loading, unloading, storage (due to vapor expansion and contraction from atmospheric conditions), and spillage. Generalization of emissions of VOCs from fuel storage and transportation is difficult, as emission factors depend on various factors including (National Pollutant Inventory, 2012):

- Type of storage tank: fixed roof tanks, external or internal floating roof tanks, pressure tanks, or variable vapor space tanks
- Climate
- Composition of the fuel
- Vapor recovery equipment

Nitrogen Oxide Emissions

Nitrogen oxides, NO_x, are a group of highly reactive gasses that include nitrogen dioxide (NO_2) and nitric oxide (NO) (Lashgari et al., 2013). Nitrogen oxides are primarily released from fossil fuel combustion and biomass burning, but natural sources such as lightning, ammonia oxidation, and soil emissions also contribute a small fraction to total emissions (Delmas et al., 1997).

Nitrogen oxides are particularly harmful to human health, causing a variety of respiratory effects and can even lead to death at high concentrations (Table 4.13). Nitrogen dioxide is of primary concern to human health as NO readily reacts in the atmosphere to form NO_2. In addition to direct health effects, nitrogen oxides can react to form other pollutants that affect air quality. Nitrogen oxides can react with VOCs to form ground

Table 4.13 Health and environmental effects of nitrogen oxides and its reaction products (Lashgari et al., 2013; Queensland Government, 2011)

Type of impact	Effects
Human health	Shortness of breath, irritation to eyes, nose, and lungs, nausea, visual impairment, olfactory impairment, reduced oxygen intake, asthma, pulmonary edema, death
Environmental	Water quality deterioration and increased nitrogen loading. See Tables 4.10 and 4.12 for effects of ozone and acid rain

level ozone and contribute to photochemical smog. Additionally, nitrogen dioxide can react with water and oxygen in the atmosphere to form nitric acid, a component of acid rain, according to the following equation:

$$4 \, NO_2 + 2 \, H_2O + O_2 \rightarrow 4 \, HNO_3$$

The health and environmental effects of ozone and acid rain are discussed previously in the sections on VOCs and sulfur dioxide emissions, respectively.

The major sources of NO_x emissions from mining activities are vehicle exhaust (see Section on SO_2) and blasting. Blasting operations emit NO_x directly to the atmosphere during surface mining operations. Typical explosives, such as ammonium nitrate/fuel oil, often become contaminated with water and drill cuttings prior to detonation, breaking down some of the ammonium nitrate (Lashgari et al., 2013). As a result, non-ideal explosive conditions occur and NO_x species (NO_2, NO) and CO form according to the following reactions depending on if the explosive is under or over fueled (Attalla et al., 2008; Lashgari et al., 2013):

$2 \, NH_4NO_3 + CH_2 \rightarrow 2 \, N_2 + CO + 5 \, H_2O$ (for over fueling/negative oxygen balance)

$5 \, NH_4NO_3 + CH_2 \rightarrow 4 \, N_2 + 2 \, NO + CO_2 + 7 \, H_2O$ (for under fueling/positive oxygen balance)

where unstable NO quickly reacts in the atmosphere to become NO_2:

$$2 \, NO + O_2 \rightarrow 2 \, NO_2$$

Emissions of NO_x from blasting at surface mines only account for a fraction of an operation's total NO_x emissions; however, the quick release and high concentrations that result may pose human health risks (Attalla et al., 2008). There have been reports of large clouds tinted brown or orange from high levels of NO_2 as a result of some large blasting operations, a potential hazard to human health (Lashgari et al., 2013; Mainiero et al., 2007).

Nitrogen oxide emissions from vehicular emissions and general mining and mineral processing can be calculated using standard emission factors.

$$NOE_{process} = \left(NOEF_{process} \right) \left(Unit_{process} \right)$$

where:

$NOE_{process}$ = nitrogen oxide emissions for a given process; lbs or kg

$NOEF_{process}$ = nitrogen oxide emission factor for given process; lbs unit^{-1} or kg unit^{-1}

$Unit_{process}$ = units for the given process; tons processed, tons produced, ton charged, tons transferred, miles traveled, etc.

Nitrogen oxide emission factors for select mining processes can be found in Appendix A, Tables A1 and A2. Emission factors in Appendix A are for uncontrolled emissions. If emission control equipment is in place, they must be taken into account when calculating emissions.

Predicting NO_x emissions from blasting is more difficult due to differences in blast design, rock types, and explosives contamination. A standard emission factor of 5.8 L NO_x/kg of ammonium nitrate/fuel oil is suggested by the U.S. EPA, but it does not take contamination into consideration (Lashgari et al., 2013).

Mercury Emissions

The release of mercury into the environment from mining and mineral processing activities represents a global threat to human and environmental health. Elemental mercury is volatile and can stay in the atmosphere for up to a year, being transported and deposited around the world. Elemental mercury is eventually deposited in the sediments of bodies of water where it is converted to methylmercury and absorbed by phytoplankton, whereby entering the food chain (World Health Organization, 2007). Once in the food chain, methylmercury bioaccumulates in higher trophic level predator species including large fish, birds, and mammals living in association with freshwater ecosystems (U.S. Environmental Protection Agency (U.S. EPA), 1997b). Mercury negatively impacts both flora and fauna (Table 4.14), and in general, methylmercury is the most toxic form to wildlife.

Human exposure to mercury typically results from direct contact with mercury vapors or eating contaminated seafood. Mercury in both elemental and methylmercury forms is toxic to the nervous system. Mercury vapors are known to cause tremors, memory loss, neuromuscular effects, motor and cognitive dysfunction, and kidney, immune, digestive, and lung effects (Table 4.14) (World Health Organization, 2007). Exposure to mercury in the womb is known to cause mental retardation, seizures, delayed development, vision and hearing loss, and language disorders (World Health Organization, 2007).

Although the large-scale use of mercury in gold mining in the U.S. has stopped, the practice is still widespread in artisanal and small-scale gold mining (ASGM). In ASGM, mercury is used to form an amalgam with

Table 4.14 Ecological impacts of mercury exposure (U.S. Environmental Protection Agency (U.S. EPA), 1997b)

Organism type	Effects of mercury
Aquatic plants	Death, growth inhibition, discoloration of leaves, leaf and root necrosis
Terrestrial plants	Death, growth inhibition, root damage, hampered nutrient uptake, reduced photosynthesis
Birds	Death, impaired growth, impaired immune response, falling, muscular incoordination, reduced reproductive success, liver, kidney, neurological damage
Fish and aquatic invertebrates	Death, impaired growth, behavioral abnormalities, reduced reproduction, reduced feeding and predatory success, emaciation, brain lesions, cataracts
Mammals	Death, weight loss, muscular atrophy, vomiting, hemorrhaging, neurological damage, heart, lung, liver, kidney, stomach damage

gold, effectively separating gold from the other materials. Mercury can be added to the whole ore (whole ore amalgamation) or to a gravity ore concentrate. The amalgam is then heated, often using a heat source such as a blow torch in open air, until the mercury is vaporized and released to the atmosphere and only the gold remains. Mercury can off-gas throughout the gold recovery process; when mercury is added to extract the gold, during the vaporization of mercury from the amalgam, and when the gold product is melted to produced doré (Telmer and Veiga, 2008). Depending on the exact process used, it is estimated that the ratio of kilograms of mercury released per kilogram of gold recovered can be anywhere from 0.001:1 up to 3:1 (Spiegel and Veiga, 2010). Operations that utilize mercury emission control and recovery technologies, such as retorts, account for the 0.001:1 ratio, while operations that use mercury with whole ore and ore concentrates have ratios of 3:1 and 1:1, respectively (Spiegel and Veiga, 2010).

Mercury amalgamation is still used in ASGM because it is one of the simplest, easiest, cheapest, and most reliable methods for gold extraction, requiring only one person, and is effective under field conditions (Telmer and Veiga, 2008). Artisanal and small-scale gold mining operations produce nearly 20–30% of the world's supply of gold, and are located in over 55 countries (Argonne National Laboratory, 2008). Such operations are often in rural areas without access to advanced processing equipment or proper education regarding the dangers mercury poses to human health, leading to the widespread use and emission of mercury in gold mining and processing.

In addition to ASGM operations, mercury may be naturally present in low-grade ore and can be produced as a by-product of gold and silver production. Low-grade ore deposits can be processed using cyanidation, a process that uses cyanide solution to extract gold or silver from the ore. Along with gold and silver, cyanidation extracts roughly 10–30% of the naturally occurring mercury contained in the ore (Mine Safety and Health Administration (MSHA), 1997). The reactions for gold and mercury extraction are as follows (Mine Safety and Health Administration (MSHA), 1997):

$$\text{Gold:} \quad 2\,\text{Au} + 4\,\text{CN}^- + \text{O}_2 + 2\,\text{H}_2\text{O} \rightarrow 2\,\text{Au(CN)}_2^- + 2\,\text{OH}^- + \text{H}_2\text{O}_2$$
$$2\,\text{Au} + 4\,\text{CN}^- + \text{H}_2\text{O}_2 \rightarrow 2\,\text{Au(CN)}_2^- + 2\,\text{OH}^-$$
$$\text{Mercury:} \quad \text{Hg}^{2+} + 4\,\text{CN}^- \rightarrow \text{Hg(CN)}_4^{2-}$$
$$2\,\text{Hg} + 8\,\text{CN}^- + \text{O}_2 + 2\,\text{H}_2\text{O} \rightarrow 2\,\text{Hg(CN)}_4^{2-} + 4\,\text{OH}^-$$

The resulting solution is concentrated and mercury vapor is released during final gold or silver recovery processes. Mercury vapor during the retorting process is captured to prevent emissions, purified, and stored; however, anytime mercury vapor is present there is a risk of atmospheric release.

Mercury emissions can also occur at many points in mineral processing, including concentrating, smelting, and leaching. Mercury is naturally present as an impurity in copper, zinc, nickel, and lead ore. During the smelting process, large amounts of mercury can be released into the atmosphere via flue gas (Pirrone et al., 2010). The amount of mercury released is highly dependent on the combustion temperatures used in boilers, roasters, and furnaces during the smelting process (Pirrone et al., 2010). In developed countries, atmospheric release is often mitigated by air pollution control devices; however, the massive scale of smelting operations means that significant emissions still occur (Wu et al., 2012). In developing countries, the lack of regulation or enforcement as well as growing trends in alternative production processes have resulted in increased atmospheric releases of mercury from the smelting of nonferrous metals (Pirrone et al., 2010).

Additionally, nonpoint source elemental mercury emissions have been measured from gold mines. Mining operations bring large amounts of mercury containing ore to the surface and expose them to the atmosphere, allowing natural volatilization. Mercury emissions have been measured from waste rock piles, heap leach pads, stockpiles, tailings, and open pits under ambient conditions (Eckley et al., 2011). In one study by Eckley et al. (2011), mercury emissions under ambient conditions at a Nevada gold mine, anywhere from <1500 to 684,000 ng m^{-2} day^{-1}, were found to be orders of magnitude greater than emissions from the natural surrounding areas. Rates of mercury emission are highly variable and depend on the following factors:

- Mercury concentration
- Moisture content
- Grain size
- Temperature
- Exposure to sun and rain
- Ore processing method

Generally, ore with higher mercury and moisture content and that is being actively processed has higher rates of mercury off-gassing (Eckley et al., 2011).

The United Nations Environment Programme (2013) had identified artisanal and small-scale gold mining (ASGM) as the largest source of anthropological emissions of mercury into the atmosphere at 37% of total global emissions, or approximately 727 tons in 2010 (Table 4.15). Other mining, smelting, and production processes for metals also represent a large portion, almost 18% or 347.5 tons, of global mercury emissions. The rapid economic growth in Asia and South America, along with the high price of

Table 4.15 Global emissions of mercury by sector in 2010 (United Nations Environment Programme, 2013)

Sector	Mercury emissions (metric tons)	Global contribution (%)
By-product or unintentional release		
Fossil fuel burning	483.9	25
Mining, smelting, and production of metals		
Ferrous metal production	45.5	2
Non-ferrous metal production	193	10
Large-scale gold production	97.3	5
Mine production of mercury	11.7	<1
Cement production	173	9
Oil refining	16	1
Contaminated sites	82.5	4
Intentional uses		
Artisanal and small-scale gold mining	727	37
Chlor-alkali industry	28.4	1
Consumer product waste	95.6	5
Dental amalgams	3.6	<1
Total	**1957.5**	**100**

gold, has led to the expansion of ASGM around the world. Effectively managing and reducing mercury emissions from ASGM operations continues to be a challenge, as the sector is typically unregulated, widespread, and mining operations are sometimes illegal and unpermitted (United Nations Environment Programme, 2013).

Mercury emissions are calculated using the following formulas (Intergovernmental Panel on Climate Change, 2006):

$$HgE_{process} = \left(HgEF_{process}\right)\left(Unit_{process}\right)$$

where:

$HgE_{process}$ = mercury emissions for a given process; g

$HgEF_{process}$ = mercury emission factor for given process; g $unit^{-1}$

$Units_{process}$ = units for given process; Mg produced, g mined, Mg combusted

Mercury emission factors for select mining and mineral processing operations are provided in Appendix A, Table A5.

A comprehensive review of over 60 studies across 19 countries concerning the health impacts due to mercury exposure from ASGM indicated that ASGM workers, their families, and children who live nearby are exposed to dangerous levels of mercury vapor, as indicated in urinary mercury concentrations (Gibb and O'Leary, 2014). Levels of mercury in those living in ASGM communities were above World Health Organization health guidance values and were high enough to where some residents experienced neurological, kidney, and autoimmune effects (Gibb and O'Leary, 2014). Additionally, the study showed that populations not directly involved in ASGM operations, specifically those that lived downstream of ASGM operations, also showed signed of neurological effects due to mercury exposure, most likely due to the consumption of contaminated fish (Gibb and O'Leary, 2014).

Particulate Matter Emissions

Particulate matter is a complex mixture of small liquid droplets and solid particulates suspended in the air. Particulate matter can originate from natural (e.g., volcanoes, fires, dust storms) or manmade sources (e.g., industrial processes, combustion, vehicle emissions). Depending on the source, particulate matter emissions may contain a variety of chemical constituents (Table 4.16).

Particles sizes are categorized according to aerodynamic diameter as less than 10 μm (PM_{10}) and less than 2.5 μm ($PM_{2.5}$), based on their respective health impacts (World Health Organization, 2013). In many places, $PM_{2.5}$

Table 4.16 Chemical constituents of particulate matter (World Health Organization, 2013)

Category	Constituent
Minerals	Silica, asbestos
Acids	Nitrates, sulfates, ammonium
Inorganic ions	Sodium, potassium, calcium, magnesium, chloride
Metals	Cadmium, copper, nickel, vanadium, zinc, lead, mercury, cobalt, chromium, arsenic, antimony, selenium
Biological components	Allergens, microbial compounds
Other	Organic and elemental carbon, particle-bound water, polycyclic aromatic hydrocarbons (PAH), soils

can constitute up to 50–70% of PM_{10} emissions (World Health Organization, 2013). Particulate emissions can be subject to long-range transportation in the air, as particles with diameters from 0.1–1.0 μm can remain airborne for days or weeks (World Health Organization, 2013).

The health effects of exposure to particulate matter have been well studied (Table 4.17). In general, $PM_{2.5}$ poses a much greater hazard than PM_{10}. Generally, PM_{10} is filtered out in the nose and upper airway and has a much shorter suspension life than $PM_{2.5}$ (Anderson et al., 2012). Due to the smaller particle size, $PM_{2.5}$ has the ability to become deeply lodged in

Table 4.17 Impacts of particulate matter on human health and the environment (Anderson et al., 2012; Gautam et al., 2012; U.S. Environmental Protection Agency (U.S. EPA), 2014e; World Health Organization, 2010)

Type of impact	Effects
Human health	Increased mortality from cardiovascular and respiratory diseases, increased respiratory and cardiovascular morbidity, aggravation of asthma, coughing, wheezing, difficulty breathing, shortness of breath, irreversible decreased in lung function, reduced lung growth rate, chronic bronchitis, silicosis, asbestosis, black lungs, berylliosis, bauxite fibrosis, increased chance of lung cancer, heart disease, and heart failure
Environment	Reduced visibility (haze), increased acidity of lakes and streams changes in coastal waters and river basins nutrient balance, reduced levels of nutrients in soil, damage to forests and crops, reduced ecosystem diversity, damage to stone and other materials

the lungs. Inhalation exposure to both PM_{10} and $PM_{2.5}$, over the short and long term, can result in negative respiratory and cardiovascular impacts, including increased mortality (Table 4.17); however, exposure to $PM_{2.5}$ is a much stronger risk factor than PM_{10} (World Health Organization, 2013). These health impacts are especially pronounced in the elderly, asthmatics, and young children. Mine workers are especially at risk of a variety of health effects (e.g., silicosis, asbestosis, black lungs, berylliosis, bauxite fibrosis) due to their close proximity to the PM emissions associated with mining operations and the hazardous compounds often released as PM from mined ore. The World Health Organization estimates that air pollution from particulate matter is the 13th leading cause of mortality worldwide, contributing to over 3.1 million deaths per year and accounting for 3% of cardiopulmonary and 5% of lung cancer deaths (Anderson et al., 2012; World Health Organization, 2010). They also estimate that exposure to $PM_{2.5}$ reduces life expectancy by up to 8.6 months (World Health Organization, 2013). There is no safe level of exposure or threshold level at which no negative effects will occur (World Health Organization, 2013).

In the environment, particulate matter, primarily $PM_{2.5}$, is responsible for reduced visibility and haze associated with many developed cities (Table 4.17). Furthermore, due to its varied chemical nature, PM can act as a pathway for nutrient transfer, upsetting delicate ecosystems, and can react in the atmosphere to form acid rain. In addition to environmental and human health risks, particulate matter can cause maintenance issues with mining equipment.

Emissions of PM from mining activities depend on the type of mine, mine location, geology of the area, terrain, operating procedures, mining equipment, vegetation, precipitation, moisture content, wind speeds, temperature, and any controls in place to mitigate emissions (U.S. Environmental Protection Agency (U.S. EPA), 1995d). Particulate matter emissions are of major concern in surface mining operations, as mining processes take place in open air. During mining and mineral processing, a majority of PM emissions are due to overburden removal, drilling, blasting, loading, unloading, transportation, wind erosion, and vehicle exhaust (Gautam et al., 2012). Overburden removal, drilling, and blasting disturb massive quantities of material, all of which has to be loaded and transported, having the potential for highly concentrated PM emissions. Blasting typically results in short-term PM emission, but at extremely high concentrations. During loading and unloading, PM emissions can be severe, depending on the wind, moisture content, and height of loading and unloading. Between loading and unloading, raw ore or coal can be temporarily stored in large piles, which are subject to wind erosion. Additional mineral processing such as crushing, conveying, and screening also contribute to PM emissions. Because mining often takes place in remote areas, frequent truck travel is required for the transportation of consumables to the mine and product from the mine. During transportation, truck tires will kick up dust and some trucks, especially coal trucks, can emit dust and PM directly from their tires, bodies, and beds and can transfer dust from unpaved roads onto paved roads that may later become suspended (Aneja et al., 2012). This can be especially problematic when driving through communities or close to homes. Throughout surface mining operations until mine reclamation, large erodible surfaces are exposed to the atmosphere, generating additional PM emissions from wind erosion. Furthermore, metal processing involves many high temperature steps such as roasting, smelting, and refining, all of which have potential for particulate matter emissions.

Particulate matter measurements taken at, or nearby, coal-mining operations show that PM emissions vary greatly depending on the region and operation (Table 4.18). To reduce human health impacts, mining laws in some areas require buffer zones between surface mining operations and populated areas; thus, the effects of PM exposure may be lessened or infrequent (Aneja et al., 2012).

Particulate matter emissions from vehicles and typical mining processes can be estimated using the following equation:

$$PME_{process} = \left(PMEF_{process}\right)\left(Unit_{process}\right)\left(1 - \frac{EC}{100}\right)$$

where:

$PME_{process}$ = particulate matter emissions for a given process, lbs or kg

$PMEF_{process}$ = particulate matter emission factor for given process; lbs unit^{-1} or kg unit^{-1}

$Unit_{process}$ = tons processed, tons produced, ton charged, tons transferred, miles traveled, kg solvent used, etc.

EC = emission control factor, %

Non-specific emissions factors for PM_{10} and $PM_{2.5}$ can be found in Appendix A, Tables A1–A2 and A6–A8; however, while emission factors for PM_{10} are widely available in government documentation, there is currently limited documentation concerning $PM_{2.5}$ emissions from mining. General emission control factors for dust control equipment is available for

Table 4.18 Particulate matter values for areas in or near surface coal mines around the world (Aneja et al., 2012)

Region	Time frame or process	PM$_{2.5}$ (μg m^{-3})	PM$_{10}$ (μg m^{-3})
North East England	Average	—	22.1
Czech Republic	Heating period	—	37
	Non-heating period	—	26
	Transition period	—	33
	Annual mean	—	33.5
Western Turkey	Drilling	—	3080
	Coal handling plant	—	1840
	Stock yard	—	1670
	Overburden loading	—	1350
	Coal loading	—	1300
Zonguldak City, Turkey	Winter	34.17	63.59
	Spring	29.84	59.16
	Summer	25.03	41.83
	Autumn	23.03	39.66
Dhanbad, India	Average	—	194 ± 32
Virginia, USA (nearby homes and public roads)	Site 1	—	250.2 ± 135.0
	Site 2	—	144.8 ± 60.0

select coal mining and general mining operations in Appendix A, Table A9. Nonspecific emission factors are helpful for estimating environmental impacts; however, if more information about mining operations is available, mine specific emission factors can be calculated and should be used if possible. Equations for estimating emission factors for some coal mining operations are available in Appendix A, Table A8.

Lead Particulates

Emissions of lead particulates from mining and mineral processing remain a problem throughout the world. Lead particulates can be transported long distances through the air and deposited via rain onto soils. Lead has no known biological function and has a broad range of toxic effects on most organisms (Table 4.19). Lead exposure is particularly dangerous for children, due to their increased exposure per unit body weight, higher physiological uptake rates, and rapid growth and development (Tong et al., 2000). Furthermore, recent evidence indicates that many of the health effects associated with children

Table 4.19 Health effects of lead exposure in selected populations (Agency for Toxic Substances and Disease Registry, 2010; Canadian Council of Ministers of the Environment, 1999; National Toxicity Program, 2012; Pourrut et al., 2011)

Population	Effects of lead exposure
Children	Decreased IQ, increased behavioral and attention-related problems, delayed puberty, decreased hearing, reduced postnatal growth, anemia, coma, convulsions, stupor, death
Adults	Decreased kidney function, reduced fetal growth in mothers, increased blood pressure, decreased fertility, nerve disorders, muscle and joint pain, cataracts, tremors, memory problems
Plants	Inhibition of germination, root elongation, development and growth, transpiration, chlorophyll production, impaired nutrient uptake, phytotoxicity
Aquatic organisms	Increased mortality, decreased abundance and diversity, abnormal development, reduced mobility

are irreversible, even with the cessation of lead exposure (Tong et al., 2000). Many of the effects of exposure to children, such as lowered IQ, and learning, behavioral, and hearing problems, can have lifelong impacts.

In the United States, annual lead emissions amounted to 743 metric tons in 2011, of which industrial processes accounted for 210 metric tons (U.S. Environmental Protection Agency (U.S. EPA), 2015c). Non-ferrous metals and ferrous metals ranked first and third, for total industrial emissions at 73 and 50 metric tons, respectively (Figure 4.5). General mining operations accounted for only 5.4 metric tons of emissions.

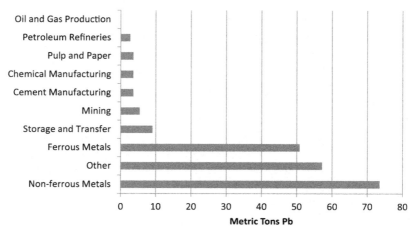

Figure 4.5 Lead emissions from industrial processes in the United States in 2011 by sector (U.S. Environmental Protection Agency (U.S. EPA), 2015c).

Table 4.20 Lead content in various types of ore (U.S. Environmental Protection Agency (U.S. EPA), 1995a)

Ore type	Lead content (%wt)
Lead	5.1
Zinc	0.2
Copper	0.2
Lead–Zinc	2.0
Copper–Lead	2.0
Copper–Zinc	0.2
Copper–Lead–Zinc	2.0

Lead in naturally found in many metallic ores and contribute to emissions during ore and metal processing (Table 4.20). The leading causes of lead emissions in mining and mineral processing in the United States are lead smelters, primary and secondary copper smelting, and iron and steel foundries. Lead can originate from nearly every smelting and refinery process including: milling, crushing, loading and unloading, sintering gasses, blast furnace gasses, and fumes from molten lead (U.S. Environmental Protection Agency (U.S. EPA), 1995c). Lead emissions from mining and mineral processing can be calculated using standard emission factors.

$$LE_{process} = \left(LEF_{process}\right)\left(Unit_{process}\right)$$

where:

$LE_{process}$ = lead emissions for a given process; lbs

$LEF_{process}$ = lead emission factor for given process; lbs unit^{-1}

$Unit_{process}$ = units for the given process; tons processed, tons produced, ton charged, tons transferred, miles traveled, etc.

Lead emission factors for select mining processes can be found in Appendix A, Tables A1 and A10. Emission factors in Appendix A are for uncontrolled emissions. If emission control equipment is in place, they must be taken into account when calculating lead emissions.

Case Study: Lead Poisoning from Gold Mining in Nigeria's Zamfara State

The Zamfara State in northern Nigeria contains significant gold deposits and the rising price of gold caused several ASGM operations spring up in the area. However, unbeknownst to the ASGM workers, the gold deposits in the area also contained high concentrations of lead. In 2010, a mass outbreak of childhood lead poisoning occurred in Nigeria as a result of ASGM activities.

During the processing of the gold ore, workers would crush and grind the ore, often at home, creating significant amounts of dust and particulates that would contaminate homes, clothes, food, and water resources (Centers for Disease Control and Prevention, 2013; Human Rights Watch, 2011). In other instances, no personal protective equipment was worn and lead contaminated dust would be brought back to the workers' homes on clothing, exposing their families and children. Furthermore, children often helped with the grinding and crushing of the raw ore, further exposing them to the lead dust (Centers for Disease Control and Prevention, 2013). In this case, children and workers were exposed to toxic levels of lead dust via inhalation and the ingestion of contaminated food and water resources (Human Rights Watch, 2011). Soil-lead concentrations in the households of some affected villages ranged from 3300 to 8000 ppm, where the U.S. EPA defines a soil–lead hazard as 400 ppm and 1200 ppm for children play areas and other parts of the yard, respectively (Dooyema et al., 2011). In two of the affected villages, all of the children under five that were tested had lead poisoning, and 97% had levels that required emergency medical intervention (Dooyema et al., 2011).

The lead outbreak, considered by the Human Rights Watch and the Centers for Disease Control and Prevention to be one of the largest and worst in modern history, killed at least 400 children and affected more than 3500 children (Centers for Disease Control and Prevention, 2013; Human Rights Watch, 2011). Incidents such as this are extremely rare, as lead poisoning is not usually associated with gold mining operations. However, this incident highlighted the lack of proper education regarding the risks of mining and mineral processing in ASGM operations. Simple measures such as the use of personal protective equipment and processing gold ore away from the household could have mitigated childhood exposure to lead dust.

Modeling Air Quality Impacts

Research concerning the reduction and mitigation of particulate and dust emissions from mining and mineral processing is ongoing. One available tool for investigating the dispersion and generation of particulates and dust is modeling. Modeling particulate dispersion is helpful for determining the potential hazards to nearby areas associated with the emission and the effectiveness of mitigation measures. Using available models, air quality impacts can be predicted to assist in:

- Determining impacts of a proposed activity such as a mine or smelter
- Designing chimney heights or emission control systems

- Ranking emission sources in terms of priority for applying emission controls
- Analyzing past air quality events
- Estimating emission rates

Estimating emissions and air quality using modeling is especially important as the U.S. EPA requires air quality analyses for air quality permit applications, all of which are based on air quality modeling (Reed, 2005; U.S. Environmental Protection Agency (U.S. EPA), 2014f). If the results of air quality modeling indicate that a facility will cause air quality to violate the National Ambient Air Quality Standards, the permit may be denied.

Air quality impacts and the effectiveness of control equipment can be compared using various modeling methods, ranging from simple to very complex. A comparative analysis to eliminate inappropriate alternative controls may use simple models, such as the box model, while thorough analyses for permit applications should use U.S. EPA approved models. There are numerous models concerning the dispersion of general pollutants; however, there are still relatively few that are specific to mining operations. A list of available models for dust and particulate dispersion from mining operations is provided in Table 4.21. Most models are not approved by the U.S. EPA.

Model complexities increase as dispersion modeling becomes non-steady state, such as in CALPUFF, a non-steady state air dispersion model preferred by the U.S. EPA. Due to increased accessibility, these complex models are becoming more widely used and increasingly varied, as models are modified with additional terms or statistical functions to more accurately represent site specific factors (Leading Practice Sustainable Development Program for the Mining Industry, 2009; Reed, 2005).

Four mathematical algorithms are used as a basis for modeling dust dispersion from mining operations:
- Box model
- Gaussian model
- Lagrangian model
- Eulerian model

Box Model

The box model is the most basic air quality modeling algorithm. The box model represents the airshed as a simple box with the contents being of

Table 4.21 Available models for dust and particulate dispersion from mining operations (Gautam et al, 2012; Reed, 2005; U.S. Environmental Protection Agency (U.S. EPA), 2014f)

Model	Type	Tested at mine sites	EPA preferred/Recommended
Surface mining			
AERMOD	Gaussian	Yes	Yes
CALPUFF	Lagrangian	No	Yes
Industrial Source Complex 3 (ISC3)	Gaussian	Yes	No
Fugitive Dust Model	Gaussian	Yes	No
Dynamic Component Program	Gaussian	Yes	No
Pereira, Soares, and Branquinho	Gaussian	No	No
Kumar and Bhandari	Eulerian	No	No
Kalgoorlie Consolidated Mines model	Unknown	Unknown	No
3D Galerkin finite-element model (FEM)	Eulerian	No	No
Herwehe finite-element	Eulerian	No	No
Fabrick	Gaussian	No	No
Winges	Gaussian	No	No
Shearer	Box	Yes	No
Underground			
Srinivasa	Eulerian	Yes	No
Wala computational fluid dynamics (CFD)	Eulerian	Yes	No
Chiang and Peng	Gaussian	Yes	No
Bhaskar	Eulerian	Yes	No
Courtney, Cheng, and Divers modified	Eulerian/Gaussian	Yes	No
Courtney, Cheng, and Divers	Eulerian	Yes	No
Courtney, Kost, and Colinet	Gaussian	Yes	No
Dawes and Slack	Gaussian	Yes	No
Hwang, Singer, and Hartz	Eulerian/Gaussian	No	No

homogeneous concentration (Reed, 2005). The model generally used is as follows (Collett and Oduyemi, 1997; Reed, 2005):

$$\frac{dCV}{dt} = QA + uC_{in}WH - uCWH$$

where:

C = concentration of pollutant throughout the box, μg m^{-3}
C_{in} = pollutant concentration entering the box, μg m^{-3}
Q = pollutant emission rate from source per unit area, μg m^{-2} s^{-1}
V = volume of the box, m^3
A = horizontal area (L × W), m^2
W = box width, m
L = box length, m
H = box height (mixing height), m
u = wind speed normal to the box, m s^{-1}

Dimensions of the box are chosen based on average wind speeds, physical aspects of the terrain, pollutant source configuration, and inversion heights. The box model is limited by the key assumption that the concentration within the box is homogeneous and fails to account for spatial information (Reed, 2005). However, it is still useful for general estimations of average pollution levels.

Gaussian Model

Gaussian models for dispersion assume that pollutant dispersion follows normal statistical distribution. Gaussian models are typically used for modeling dispersion from buoyant air pollution plumes. The model generally used is as follows (Reed, 2005):

$$X = \frac{Q}{2\pi u_s \sigma_y \sigma_z} \left[\exp\left\{ -0.5\left(\frac{y}{\sigma_y}\right)^2 \right\} \right] \left[\exp\left\{ -0.5\left(\frac{H}{\sigma_z}\right)^2 \right\} \right]$$

where:

X = hourly concentration at downwind distance x, μg m^{-3}
u_s = mean wind speed at pollutant release height, m s^{-1}
Q = pollutant emission rate, μg s^{-1}
σ_y = standard deviation of lateral concentration distribution
σ_z = standard deviation of vertical concentration distribution
H = pollutant release height (stack height), m
y = crosswind distance from source to receptor, m

The terms σ_y and σ_z are based on atmospheric stability coefficients, where larger values (usually at greater distances from the source) represent a plume with wide spread and low peak, and vice versa (Reed, 2005). Pollutant release height, H, may vary due to the vertical velocity of gas leaving the stack and buoyancy as warm stack gases rise in the cooler surrounding atmosphere.

Gaussian models, while the most commonly used, are not without limitations. The model assumes that wind speed and direction is constant, emission rates are constant, the terrain is flat, deposition is negligible, and the shape of the plume is conical (Reed, 2005). Regardless, Gaussian models have proven to be accurate within 20% at ground level at distances less than 1 km, and accurate within 40% for elevated emissions (Reed, 2005).

Lagrangian Model

The Lagrangian model calculates dispersion of plume parcels using a shifting reference grid based on the wind direction and the general direction of plume movement (Reed, 2005). The motion of plume is modeled based on a random walk process and the reference grid follows the plume. The Lagrangian model is most notably used in the CALPUFF model, one of the few models recommended by the U.S. EPA. The model generally used is as follows (Collett and Oduyemi, 1997; Reed, 2005):

$$\langle c(r,t) \rangle = \int \int_{-\infty}^{t} p(r,t|r',t')S(r',t')dr'dt'$$

where:

$\langle c(r,t) \rangle$ = average pollutant concentration at location r at time t

$p(r,t|r',t')$ = probability function that an air volume is moving from location r' at time t' to location r at time t.

$S(r',t')$ = source emission term

The probability function is derived from site specific meteorology, particle size distribution, and particle density. The Lagrangian model is limited due to its dynamic nature. By using a moving reference grid, direct comparisons between actual field measurements, usually taken at fixed points, is difficult (Reed, 2005). A fixed Eulerian grid can be used to facilitate direct comparisons.

Eulerian Model

The Eulerian model is similar to the Lagrangian model with the exception that it utilizes a fixed reference grid, as opposed to the moving grid of the Lagrangian model. Both models track the movement of pollution plumes over time, but the Eulerian model observes a fixed grid as the plume passes by. The Eulerian model follows the general form (Collett and Oduyemi, 1997; Reed, 2005):

$$\frac{\partial \langle c_i \rangle}{\partial t} = -\overline{U} \cdot \nabla \langle c_i \rangle - \nabla \cdot \langle c_i' U' \rangle + D \nabla^2 \langle c_i \rangle + \langle S_i \rangle$$

where:

U = windfield vector $U(x,y,z)$, $U = \overline{U} + U'$

$\overline{U}$ = average wind field vector

U' = fluctuating wind field vector

c_i = concentration of pollutant for ith species, $c = \langle c \rangle + c'$

$\langle c \rangle$ = average pollutant concentration, where $\langle \ \rangle$ denotes average

c' = fluctuating pollutant concentration

D = molecular diffusivity

S_i = source or sink term (chemical reactions should be taken into account)

The terms $\overline{U} \cdot \nabla \langle c_i \rangle$, $\nabla \cdot \langle c_i' U' \rangle$, and $D \nabla^2 \langle c_i \rangle$ represent the rates of advection, turbulent diffusion, and molecular diffusion, respectively. For most cases, the wind field vector U is considered to be turbulent and requires average and fluctuating wind field vector components. This has a similar effect on the pollutant concentration, c_i. The Eulerian model can be difficult and computationally expensive to solve in three dimensions, thus, it is sometimes reduced to one or two dimensions for convenience (Reed, 2005). It is further limited in that the resolution of the model is limited to spatial and temporal grid on which it is solved (Collett and Oduyemi, 1997).

4.3 WATER QUALITY AND QUANTITY IMPACTS FROM MINING

Water is essential for sustaining life, and water of high quality is necessary for agricultural, industrial, recreational, and domestic use. Water quality can generally be defined as the physical, chemical, and biological characteristics of water relative to the needs of biotic species and for human use. Negative impacts to water quality and quantity from mining are certain to be of appreciable concern to nearby citizens (Jain et al., 2012), as it can affect regional water supplies, contaminate water resources, and destroy sensitive aquatic habitats. Water quality is vulnerable due to the interconnected nature of water resources. The hydrologic process consists of two interrelated phases: surface water and groundwater. Surface water, such as streams, lakes, or wetlands, can recharge groundwater and any impact on surface water may affect groundwater in the future and can affect water quality downstream. Groundwater in turn may sustain wells, subterranean ecosystems, and surface waters such as rivers and wetlands.

Due to the interconnectedness of water resources, water quality should be protected at the source to reduce negative impacts on nearby communities and local flora and fauna. In order to reduce water impacts, it is important to understand how mining processes contribute to negative impacts on surface water and groundwater quality and quantity and the associated environmental implications. Improving water system design and management to reduce water withdrawal and negative impacts is key to move the mining industry toward a more sustainable future. This section will cover the direct impacts of mining and mineral processing on water quantity and quality.

Water Quantity Impacts

Water is crucial to the mining industry and is used during all stages of a mine's life, from exploration to rehabilitation. Water is essential for mineral processing, metal recovery, cleaning, pumping and transportation, cooling, dust control, and worker needs. Table 4.22 summarizes the major water uses and water management considerations during each phase of a mine's life. Mining operations result in the production, use, and disposal of various types of mine water. Types of water associated with mining and mineral processing are described in Table 4.23.

The U.S. Geological Survey (Maupin et al., 2014) has estimated that mining operations in the United States withdrew approximately 20.1 million m^3 of water per day in 2010. Of this, approximately 5.3 million m^3 were from surface water sources, and the remaining 14.8 million m^3 were from groundwater. Although mining only represented about 1% of national water withdrawals in 2010, total mining withdrawals have increased 39% compared to 2005 levels, with a 54% increase in groundwater withdrawals (Maupin et al., 2014). Globally, water usage by the mining industry is estimated to be around 2–4.5% of average national water usage, even in mining intensive countries like Chile and Australia (Gunson et al., 2012). Worldwide estimates of water usage by the mining industry range from 7–9 billion m^3 per year (Global Water Intelligence, 2011). However, this still represents significant volumes worldwide, as just the copper mining

Table 4.22 Major water uses and water management considerations at various stages of mining (Leading Practice Sustainable Development Program for the Mining Industry, 2008)

1. Exploration

- Temporary water supply needed
- Impacts of water management on local water resources and users
- Potable water treatment needed
- Discharge of excess drilling water
- Wastewater disposal
- Storm water management

2. Resource development and design

- Identification and quantification of water supply
- Study of impacts of water abstraction and diversion on local water resources and users
- Government permits and approvals needed
- Water supply, storage, and treatment
- Dust suppression and dewatering disposal
- Wastewater disposal
- Storm water management

3. Mining, minerals processing, and refining

- Water supply management
- Water treatment
- Mine dewatering
- Worked water recovery, storage, reuse, and disposal
- Dust control
- Contamination management
- Catchment management
- Acid mine drainage

4. Shipping

- Spillage is possible
- Dust control necessary

5. Rehabilitation

- Post-mining landform drainage design
- Site remediation
- Water supply scheme decommissioning
- Decommissioning of facilities
- Mine pit lake modeling and closure

6. Post-mining and closure

- Rehabilitation performance monitoring
- Erosion control and drainage maintenance
- Contaminated site remediation verification

Table 4.23 Types of mine water

Type of mine water	Description
Mine water	Surface or groundwater present at the mine
Mill water	Water used from crushing and sizing ore
Process water	Water used for hydrometallurgical processes and extraction, may contain process chemicals
Mining water	Water that contacts any of the mine workings
Leachate	Mine water percolated through solid mine wastes (e.g., tailings)
Effluent	Mine, mill, process, or mining water discharged to surface waters
Mine drainage	Surface or groundwater that flows from the mine site to surrounding areas
Acid mine drainage	See Section 4.4

Adapted from Lottermoser (2010).

industry alone was estimated to withdraw over 1.3 billion m^3 of water worldwide in 2006 (Gunson et al., 2012). Water usage can vary depending on the size of the mine, commodity mined, mining methodology, the method of ore processing, and water recycling practices. Table 4.24 provides typical water usage for major water requiring processes in a 50,000 ton per day low grade copper mining operation in an arid climate. The development of new mining operations in arid regions, increased environmental and wastewater standards, and increasing demand for water requires that water quantity considerations be taken into account for all mining operations.

Table 4.24 Typical water usage for a 50,000 ton per day copper mining operation

Process	Water usage (m^3/d)
Flotation	115,000
SAG mill cooling	4100
Ball mill cooling	4100
Compressor cooling	4100
Road dust control	3500
Froth wash	2900
Pump gland seal water	1400
Reagent dilution	700
Primary crusher dust control	350
Coarse ore stockpile dust control	120
Staff domestic water	60

Adapted from Gunson et al. (2012).

Groundwater provides about 50% of the world's drinking water; there is a growing concern that mining and mineral processing activities are extracting local water resources in an unsustainable manner (National Academy of Engineering (NAE), 2010). Groundwater withdrawals for mining activities may affect *groundwater quantity* by lowering the water table elevation. Overpumping of groundwater may reduce the aquifer safe yield, defined as the groundwater supply that is available for use without ultimate depletion of the aquifer. Furthermore, changes in ground surface conditions (due to land clearing and surface mining), permeability, and porosity can increase runoff and decrease percolation, reducing groundwater recharge rates and further decreasing aquifer safe yield.

Groundwater can also be affected by mine dewatering, where pit or underground mines experience natural groundwater inflow and must continually pump out water to maintain access to the mining site. Continuous pumping in this manner will cause a local decrease in the water table surrounding the mine. In some cases, after a mine is abandoned, pumping will cease and groundwater is allowed to naturally flood the mine, which can further lower the water table. In abandoned pit mines, the natural flooding may form a lake, which can increase groundwater loss through evaporation that would not have taken place otherwise (International Council on Mining & Metals (ICMM), 2012). Regardless of the cause, lowered water tables can result in nearby wells running dry and may impact nearby *surface water quantity* (e.g., springs, lake levels, base flow in streams, and resource availability for local communities). A long-term impact of lowering the water table is subsidence, resulting in a decreased amount of future storage in the aquifer and deformation of the land surface that can damage infrastructure, buildings, and wildlife habitat (See Section 4.5 for more details on subsidence). Decreased water tables may also negatively impact local ecology, as plant root systems need to travel deeper to access groundwater and wildlife access to surface water and springs may decrease, creating conditions more conducive to select species. Water table decreases may also cause sensitive cave ecosystems to dry up. Changes in withdrawal and recharge rates can affect aquifer safe yield, limiting available water resources for nearby communities for the foreseeable future. Short-term impacts of groundwater changes are difficult to quantify due to the slow passage of groundwater through the soil.

When water is withdrawn or discharged directly to/from surface water sources, it can affect surface water flows, aquatic habitat, groundwater recharge, and water resources (Table 4.25). Other mining activities, such

Table 4.25 Environmental impacts on quantity attributes of surface water and ground water (Leading Practice Sustainable Development Program for the Mining Industry, 2007a)

Water quantity attributes	Effect on health or environment	Source
Aquifer impacts	Lowered water table can impact groundwater dependent vegetation and fauna that depend on surface water sustained by groundwater. Decreased water table can reduce aquifer safe yield and access to groundwater via wells.	Mine dewatering, water withdrawal, mine flooding, changes in ground surface properties.
Watershed alterations	Increased or decreased flows in streams can impact dependent ecosystems and communities.	Mine dewatering, diversion or damming of streams, development of interfering infrastructure.

as land clearing, dam construction, storm water management, and river diversions can alter runoff patterns and percolation rates, leading to increased surface water flows in some areas and decreased flows in others. Many species of aquatic plants and animals are sensitive to variation in flow rates and require specific flow conditions. Extreme flow conditions (i.e., high flow and low flow) present the greatest risks. Low flows can reduce habitat, nutrient and sediment transport, surface aeration, and the assimilative capacities of bodies of water, which can be especially important during summer conditions when there is typically high biological activity and dissolved oxygen is at a minimum (Jain et al., 2012). High flows can inundate nearby vegetation, erode riverbanks, cause river scouring, and damage infrastructure and mining operations (Jain et al., 2012). Extreme changes in surface water flows affect all receive bodies of water downstream and the communities that rely on them. As such, surface water withdrawals for mining are usually highly regulated by government agencies due to the many parties that rely on surface water (International Council on Mining & Metals (ICMM), 2012).

Water quantity impacts from mining are not always negative. Mining operations can sometimes become a source of water for local communities. For example, the Trekkopje uranium mine in Namibia, operated by AREVA, cannot directly utilize local groundwater resources due to high salinity levels and available freshwater resources cannot support both mining operations and the local communities (International Council on Mining & Metals (ICMM), 2012). As a result, AREVA operates a seawater desalination plant that provides water for mining operations, but at the same time provides excess water to the local community. In another case, groundwater inflow into three mining sites operated by Anglo American in South Africa is being treated and distributed to the local water-stressed community, providing up to 12% of the community's daily water needs (International Council on Mining & Metals (ICMM), 2012). In some cases, mining operations may not directly benefit from water treatment operations, such as the Cerro Verde copper and gold mining operation in Peru, where wastewater generated by the local community is treated to reduce environmental impacts, but is not intended for use by the mining operation (International Council on Mining & Metals (ICMM), 2012).

Water Quality Legislation

In order to fully understand the implications of the negative impacts on water quality from the mining and mineral processing industry, it is necessary to have a brief background in water quality standards and legislation. In the U.S., water quality is protected under the Clean Water Act (33 U.S.C. 1251 et seq.) and the Safe Drinking Water Act. Other industrialized countries have similar standards that should be followed accordingly.

The primary goal of the Clean Water Act (CWA) is to restore and maintain the physical, chemical, and biological integrity of the nation's water resources. In the U.S., CWA was originally passed in 1972 and has since been amended by the Clean Water Act of 1977 and the Water Quality Act of 1987. The discharge of pollutants via wastewater or storm water into surface waters from point sources is regulated by the CWA under the National Pollutant Discharge Elimination System (NPDES). The NPDES permit program is administered by each state and requires dischargers to obtain a permit and meet all treatment and discharge requirements (U.S. Environmental Protection Agency (U.S. EPA), 2014a). Water quality criteria may vary based on national or state standards designed to protect aquatic life, or recreational, agricultural, industrial, or public water supply use. The National Recommended Water Quality Criteria for both aquatic life and human health is maintained by the U.S. EPA and contains recommendations for nearly 150 pollutants (U.S. Environmental Protection Agency (U.S. EPA), 2014d). Nonpoint source pollution, including runoff from construction sites and acid mine drainage, is managed by states on a watershed-by-watershed basis through the implementation of best management practices (U.S. Environmental Protection Agency (U.S. EPA), 2014a).

Under the authority of the Safe Drinking Water Act, the U.S. EPA established National Primary Drinking Water Regulations for contaminants that are adverse to human health. Maximum Contaminant Levels for microorganisms, disinfectants, disinfection byproducts, inorganic and organic chemicals, and radionuclides were established. National Secondary Drinking Water Regulations are also available, but are non-enforceable guidelines to improve the cosmetic and aesthetic nature of drinking water. Primary and secondary maximum contaminant levels for select contaminants are provided in Table 4.26.

Many countries have similar national water quality standards for drinking water and environmental protection that are enforced by state or provincial agencies. Canada has several environmental quality guidelines: Canadian Water Quality Guidelines for the Protection of Aquatic Life, Recreational Water Quality Guidelines and Aesthetics, Metal Mining Effluent Regulations, and Guidelines for Canadian Drinking Water Quality. The European Union adopted the Urban Waste Water Treatment Directive, the Water Framework Directive, and the Drinking Water Directive to manage water resources and potable water quality. Member states are individually responsible for water quality and pass national legislation according to these directives. Internationally, the World Health Organization has developed similar guidelines for drinking water quality (World Health Organization, 2011).

Water Quality Impacts from Mining and Mineral Processing

Groundwater and surface water *quality* may be affected by the "handling, storage, and disposal of mining wastes; the mine excavation itself; water-table drawdown; wastewater discharge; and the storage and handling of chemicals, reagents, and fuels" (Department of Natural Resources (DNR), 2003). Negative impacts can depend on the minerals being mined, mining technology and processes being used, sensitivity of aquatic habitats and water resources, and the operator's environmental commitment and monitoring programs (Safe Drinking Water Foundation, 2005). Pollutants can be organic or inorganic. Fish kills are an immediate effect of water pollution, while long-term effects can include pollution of drinking water for downstream communities and degradation of wildlife habitats. Similarly to water quantity impacts, negative impacts on water quality can affect surface and groundwater, which in turn can impact dependent flora, fauna,

(U.S. Environmental Protection Agency (U.S. EPA), 2009a)

Contaminant	Primary/Secondary	Maximum contaminant level	Sources
Turbidity	Primary	—	Soil runoff
Antimony	Primary	0.006 mg/L	Discharge from petroleum refineries
Arsenic	Primary	0.01 mg/L	Erosion of natural deposits
Asbestos	Primary	7 million fibers per liter	Erosion of natural deposits
Barium	Primary	2 mg/L	Discharge from metal refineries
Beryllium	Primary	0.004 mg/L	Discharge from metal refineries
Cadmium	Primary	0.005 mg/L	Discharge from metal refineries
Chromium	Primary	0.1 mg/L	Erosion of natural deposits, discharge from steel mills
Cyanide	Primary	0.2 mg/L	Discharge from steel factories
Fluoride	Primary	4.0 mg/L	Discharge from aluminum factories
Copper	Primary	1.3 mg/L	Erosion of natural deposits
	Secondary	1.0 mg/L	
Lead	Primary	0.015 mg/L	Erosion of natural deposits, discharge from metal refineries
Mercury	Primary	0.002 mg/L	Erosion of natural deposits, discharge from metal refineries
Nitrate	Primary	10 mg/L	Erosion of natural deposits, runoff from fertilizers
Selenium	Primary	0.05 mg/L	Erosion of natural deposits, discharge from refineries and mines
Thallium	Primary	0.002 mg/L	Leaching from ore processing
Uranium	Primary	30 μg /L	Erosion of natural deposits
Hexachlorobenzene	Primary	0.001 mg/L	Discharge from metal refineries
Trichloroethylene	Primary	0.005 mg/L	Discharge from metal degreasing
1,1,1-Trichloroethane	Primary	0.2 mg/L	Discharge from metal degreasing
Benzene	Primary	0.005 mg/L	Leaching from gas storage tanks
Aluminum	Secondary	0.05–0.2 mg/L	—
Iron	Secondary	0.3 mg/L	—
Manganese	Secondary	0.05 mg/L	—
pH	Secondary	6.5–8.5	—
Silver	Secondary	0.1 mg/L	—
TDS	Secondary	500 mg/L	—
Zinc	Secondary	5 mg/L	—
Chloride	Secondary	250 mg/L	—

and human usage. A summary of water quality impacts from mining is found in Table 4.27.

In addition to the attributes listed in Table 4.27, mining effluent may also directly impact the pH, dissolved oxygen levels, and temperature of receiving waters. Any discharge to surface water, whether from storm water runoff, untreated process water, accidental discharge, or treated mine water, if not carefully regulated, can impact all three of these attributes. Aquatic organisms are highly sensitive to small changes in pH and temperature in their environment, which can negatively impact reproduction and growth rates, and even cause death if variations are great enough (Younger and Wolkersdorfer, 2004). Low pH waters are also more likely to contain higher concentrations of soluble heavy metals and other toxic constituents. Thermal pollution can increase the temperature of receiving waters, lowering oxygen solubility, and reducing the dissolved oxygen available for aquatic organisms. Lack of dissolved oxygen can lead to widespread fish kills and dead zones.

Mine dewatering, in addition to affecting groundwater quantity, can negatively impact surface water salinity levels if discharged without treatment. Depending on the local geology and hydrogeological conditions, groundwater pumped out of mining sites can be of relatively high quality (high enough to use directly for irrigation or other purposes with minimal treatment), or can be highly saline (Younger and Wolkersdorfer, 2004). Saline water discharged to freshwater streams can increase salinity levels to the point where freshwater species diversity and growth become negatively impacted. Negative impacts are seen in many aquatic species at salinities around 1000 mg/L (Younger and Wolkersdorfer, 2004). Tolerant species may continue to grow, but others will avoid areas with increased salinity, effectively reducing their habitat. Although they may experience negative impacts, many freshwater aquatic species can tolerate salinity levels anywhere from 1000–4000 mg/L, but freshwater biodiversity greatly decreases above 10,000 mg/L (Younger and Wolkersdorfer, 2004).

Suspended Solids

Suspended solids consist of organic or inorganic particulate matter, usually less than 62 μm but sometimes in the form of larger flocs, that is suspended in the water column by turbulence (Bilotta and Brazier, 2008). Suspended solids can consist of settleable, nonsettleable, and floating materials, such as silica, clay, organic matter, microorganisms, and silt (Jain et al., 2012). Although all streams naturally have some amount of suspended solids,

Table 4.27 Summary of environmental impacts on quality attributes of surface water and ground water (Leading Practice Sustainable Development Program for the Mining Industry, 2007b, 2008; Younger and Wolkersdorfer, 2004)

Water quality attribute	Description	Source	Effect on health or environment	Mitigations
Suspended solids	Small particles of sediment, organic matter, and inorganic matter	Erosion of disturbed land around mine (stockpiles of waste rock/ore, tailings piles and dams, roads, maintenance areas, exploration areas, reclamation areas).	Alters aquatic habitat, toxic for fish, destroys habitat for benthic organisms, loss of storage capacity in waterbodies, could lower pH of water systems, increase BOD of system.	Discharge diversions, drainage/storm water conveyance, runoff dispersion, sediment control and collection, vegetation/ soil stabilization, capping of contaminated sources, end-of-pipe treatment.
Acid mine drainage (AMD) (see Section 4.4)	Water has low pH, high concentration of heavy metals and sulfates	Water, oxygen, and metal sulfides exposed during mining accelerate a natural process, microbes that thrive in a low pH environment increase the amount of AMD created. Tailings storage facilities are high risk.	Destroys stream habitats and wildlife, can occur indefinitely, high concentrations of metals can be toxic.	Prediction of acidic regions and reduction of mining, prevent/minimize oxygen from contacting material, material with neutralizing capabilities can be applied, isolation of acid generating materials, treatment. See section on Acid Mine Drainage for more details.

Continued

Table 4.27 Summary of environmental impacts on quality attributes of surface water and ground water (Leading Practice Sustainable Development Program for the Mining Industry, 2007b, 2008; Younger and Wolkersdorfer, 2004)—cont'd

Water quality attribute	Description	Source	Effect on health or environment	Mitigations
Dissolved solids	Heavy metals, sulfates, saline intrusion	AMD, mine workings, overburden and waste rock piles, mined ore, tailings piles and processing facilities, chemical storage areas, reclamation activities, discharge of process water, mine water, runoff and seepage of these wastes.	Heavy metals are highly toxic to aquatic organisms including fish and macro-invertebrates, can be found in sediment in surface water, and can bioaccumulate in fish and the food chain. Sulfates are not generally toxic, but can cause diarrhea and affect the taste of water.	Sedimentation BMPs, preventing AMD, reducing erosion, monitoring mine dewatering.
Toxic compounds	Cyanide	Sodium cyanide used in the leaching of gold and silver. Can directly contaminate water or form poisonous hydrogen cyanide gas at a pH lower than 8 or 9.	Highly toxic to birds, fish, and aquatic invertebrates. Less toxic to aquatic plants. May contaminate drinking water downstream. Can cause nerve damage, convulsions, coma, and death in humans.	Liners and site prep required for heap leach piles and tailings dams, monitoring of solutions, treatment of wastes, reduce/eliminate access of wildlife to cyanide solutions, reduce cyanide concentrations, reclamation activities.
Nutrients	Nitrogen and phosphorus compounds	Septic and sewage waste, soil erosion, fertilizers, explosive residues, process water, process chemicals, landfill leachate.	Nutrients in streams lead to algae blooms and subsequent eutrophication, decreasing dissolved oxygen and resulting in fish kills.	Passive treatment with wetlands, active treatment with precipitation, biological treatment, aeration, or oxidation.

mining and mineral processing operations can contribute to suspended solids loading in surface water resources due to increased erosion from disturbed land around the mine, including stockpiles of waste rock or ore, tailings piles and dams, roads, land clearing, construction, open pits, and improper reclamation. All mentioned activities change surface runoff patterns and can consequently increase storm water flow. Other sources may include direct wastewater discharges, dewatering operations, and wastewater treatment effluent (Younger and Wolkersdorfer, 2004).

Suspended solids can alter the chemical and physical properties of surface waters, all of which can have significant impacts on aquatic organisms (Table 4.28) (Bilotta and Brazier, 2008). Large influxes of organic matter can rapidly decompose, depleting dissolved oxygen levels and causing massive fish kills. Suspended solids with light-scattering properties can reduce light penetration, whereby reducing primary production and algal and planktonic biomass (Younger and Wolkersdorfer, 2004). Fish that rely on algae and plankton as a food source will in turn be negatively impacted. Heavy metals, which are highly toxic to aquatic organisms, may also be released from suspended solids and waste rock originating from mining operations.

Table 4.28 Effects of suspended solids on surface water (Bilotta and Brazier, 2008; Jain et al., 2012; Younger and Wolkersdorfer, 2004)

Type of impact	Effects
Physical	• Increased turbidity and reduced light penetration • Change in water temperatures • Deposition and infilling of reservoirs and channels • Reduced navigability
Chemical	• Release of adsorbed heavy metals, pesticides, nutrients (e.g., phosphorus) • Increases biological oxygen demand; depletes dissolved oxygen
Ecological	• Reduced primary productivity • Reduced periphyton and algal biomass • Reduced invertebrate populations and diversity • Reduced fish growth and foraging behavior, damage to gills • Fish kills and increase mortality of eggs and smelt • Abrasive damage to aquatic plants and fish gills • Blanketing bottom dwelling organisms

Heavy Metals

Heavy metals are naturally found in many rocks and ore. The excavation and processing of such metal rich ore and rock provides a potent contamination pathway that can affect both surface water and groundwater quality. Heavy metals consist of a group of hazardous inorganic chemicals and include (U.S. Department of Agriculture, 2000):

- Cationic
  - Lead
  - Chromium
  - Zinc
  - Cadmium
  - Copper
  - Mercury
  - Manganese
  - Nickel
- Anionic
  - Arsenic
  - Selenium
  - Molybdenum

Heavy metals in waste rock and ore can increase in solubility and become mobilized when exposed to oxygen and water, with acidic water being of particular concern. Possible sources of heavy metal contamination of water resources from mining and mineral processing include acid mine drainage (see Section 4.4 for more details), heap leaching, in situ leaching, failure of tailings dams, increased suspended solids concentrations due to erosion, and the improper disposal of mining water and process water. Water is essential for many mineral processing and separation operations and wastewater from such operations can contain both heavy metals and toxic chemical constituents. Although water recycling has become increasingly common, in areas with less stringent mining regulations, wastewater is sometimes discharged directly to surface waters or to land where it can infiltrate and contaminate groundwater. Failure of tailings dams or seepage can cause heavy metal laden water to be discharged into surface water or to seep down and contaminate underlying groundwater. Failures can be due to a variety of reasons including poor construction and maintenance, floods and earthquakes, and ground subsidence (Bempah et al., 2013).

In heap leaching, ore is excavated, crushed, and placed on lined leach pads where a leach solution is then percolated through the ore. As the leach solution passes through the ore, it will solubilize specific minerals and metals, which can then be extracted from the pregnant solution and refined. The solution is usually recycled and reapplied in a closed loop process. The solution used depends on the target metal: cyanide is used for gold and silver; sulfuric acid is used for copper, nickel, and uranium. Heap leaching can impact water quality due to failures in maintaining a closed process such as leaks, leach pad failures, dam failures and tailings spills, chemical spills, and other unplanned discharges (Mineral Policy Center, 2000). In addition to containing heavy metals, leach solutions may contain cyanide, which is highly toxic to humans, mammals, birds, and fish, and can render water resources undrinkable (see Table 4.27) (Mineral Policy Center, 2000).

In situ leach mining of uranium poses a great risk of groundwater heavy metal contamination. In situ leach mining consists of injecting a leach solution into the ore deposit below the water table, mobilizing the target metal, and recovering and pumping the pregnant solution to the surface via production wells (International Atomic Energy Agency, 2005). Although this method limits land disturbances and minimizes waste rock and tailings, the leach solution, usually sulfuric acid, hydrofluoric acid, ammonia, ammonium bicarbonate, or sodium bicarbonate, will contain high concentrations of heavy metals, and can migrate away from the treatment areas, contaminating nearby groundwater and surface water resources (Lottermoser, 2010). The migration of leaching solution away from the mining area is minimized by maintaining a hydraulic cone of depression within the injection zone; however, significant pump and treat restoration will still be required and returning the groundwater to its original condition is difficult and sometimes impossible (International Atomic Energy Agency, 2005; Lottermoser, 2010).

Heavy metals are highly toxic to aquatic organisms and contamination of water resources can impact the drinking water of communities downstream. Furthermore, heavy metals tend to accumulate in the food chain, exposing those who regularly consume fish, plants, or animals from contaminated areas to high levels of heavy metals. Information regarding the effects of heavy metals on human health, aquatic organisms, and terrestrial organisms can be found in Table 4.35.

Modeling Water Quality and Quantity

Modeling is used throughout the mining industry as a decision support tool to assess water management techniques, mitigation of water contamination, and design water infrastructure (Golder Associates, 2011). Through modeling, water quality and quantity impacts can be predicted for a wide range of activities including the following (Leading Practice Sustainable Development Program for the Mining Industry, 2008):

- Quantity
  - Amount of water to be pumped from a mine (dewatering)
  - Diversion of water from one catchment to another
  - Altering the physical terrain and changing natural drainage, thereby creating a time shift of rainfall runoff by capture, treatment, and reducing infiltration
  - Increase/decrease of base stream flows due to discharges
  - Conversion of temporary waters to perennial waters
- Quality
  - Effects of spills, drainage, runoff, or accidental discharges
  - Effects of tailings dams and waste disposal facilities

Alternatives should be considered for mine water reduction and reuse and mitigating water quality impacts during all phases of mine construction, operation, and management. Similar to the process for evaluating alternatives for air quality, different alternatives for design and mitigations can be compared using various modeling methods, ranging from simple to very complex.

A simple mass balance can give a general idea of which alternative processes and control technologies are most applicable, though more complex mass balances can be developed depending on the constituent. Mass balances can be used for constituents such as dissolved oxygen, chloride (total dissolved solids), total phosphorus, and total organic carbon (Canter, 1996). The generic form of the mass balance equation is as follows:

$$C_{avg} = \frac{\sum C_i Q_i}{\sum Q_i}$$

where:

C_{avg} = Average concentration of constituent for combined discharges stream, mg/L

C_i = concentration of constituent in ith discharge stream, mg/L

Q_i = flow for ith discharge stream, L/s

After alternatives with the least hydrologic impacts have been chosen, complex mathematical models can be used to predict more precise impacts.

Examples of mathematical models include the classic dissolved oxygen model by Streeter-Phelps or a one dimensional advection-dispersion mass transport equation integrated over space and time for each water quality constituent (Canter, 1996). The below equation shows an example of an advection-dispersion equation.

$$\frac{\partial C}{\partial t} = -v\frac{\partial C}{\partial x} + D_L\frac{\partial^2 C}{\partial x^2} - \frac{\partial q}{\partial t}$$

where:

C = concentration of constituent in water, mg/L

v = pore water flow velocity, m/s

D_L = hydrodynamic dispersion coefficient, m^2/s

t = time, s

x = distance, m

q = concentration of constituent in the pores of the solid phase, mg/L

Many of the simpler models do not take into account the interconnectivity between an aquifer and a watershed. If an impact is determined to be a high risk, an integrated water model can be used to determine the change in impacts due to the interaction of groundwater and surface water. Integrated water models can provide a more accurate representation of reality; however, they are time intensive and require more data and interpretation. Computer programs that use these complicated mathematical models can ease prediction of impacts to water quality and quantity.

Complex mathematical models for water quality modeling can be classified according to various considerations (Canter, 1996):

• Underlying bases: deterministic or probabilistic
• Flow regime: steady state or dynamic
• Dimensional considerations: 1D, 2D, or 3D
• Type of water body: lake, river, estuarine, coastal, ocean

Deterministic models operate using input parameters based on average water quality and quantity conditions and are used to predict scenarios, such as extreme precipitation, climate change, and sensitivity analysis (Golder Associates, 2011). Probabilistic models utilize probability distributions as inputs in order to represent uncertainty in the modeled system and are used for uncertainty analyses (Golder Associates, 2011). Model selection should be determined by engineering judgment and the mine life phase being modeled.

Water and mass balance modeling should be conducted during the feasibility and planning stages of a mining operation and cover the whole life cycle of the mine, from exploration to long past reclamation. Simple deterministic models may be appropriate for feasibility studies, but more complex models may be necessary as the mine moves to the planning, construction, and operations phases (Golder Associates, 2011). Mine plans should specify the length of each phase of the mine's life and can be used to guide water and mass balance model input parameters.

Water and mass balance modeling requires multiple model inputs; a summary of inputs is provided in Table 4.29. Mine process and dewatering inputs are essential for accurate modeling of water balance and quality. Mine plans should specify timelines for operations and processing technology implementation to assist in the evaluation of mine process inputs. Physical inputs are used to establish drainage basins, runoff coefficients, and determine how topography affects the flow of water in and around the mining site (Golder Associates, 2011). Runoff coefficients should be supported by climate, physical, hydrological, and hydrogeological inputs. Climatic inputs should be determined by establishing meteorological stations at the mining site during the exploration and feasibility phases. Multiple stations may be required if the mining site encompasses a large elevation change, several valleys, or areas with unique microclimatic conditions (Golder Associates, 2011). Hydrological and hydrogeological inputs are used to determine runoff coefficients and inflows into open pits and underground mines (Golder Associates, 2011). Water quality inputs should be determined by water sampling during the feasibility and planning phases. Sampling is important to establish baseline water quality values in areas without historical data (Golder Associates, 2011). Surface water hydrometric stations and groundwater boreholes can be used to monitor and determine model inputs. Geochemical inputs include the types of rocks and minerals at the mining location and are important for determining potential impacts to water quality from water draining through stockpiles, tailings, and mine workings (Golder Associates, 2011). Accurate geochemical inputs may require static and kinetic testing using sample rock and tailings to determine drainage chemistry (Golder Associates, 2011). A list of common water quality models is provided in Table 4.30.

Table 4.29 Inputs for water and mass balance modeling for mining operations

Input category	Inputs
Mine process and dewatering	• Ore throughput • Minimum freshwater requirements • Make-up, recycled, and reclaimed water requirements • Tailings production and tailings slurry water content • Water leaving with processed ore • Irrigation rate for heap leaching • Saturated water content and residual water content after drain down for heap leaching • Water storage capacity • Potable water requirements • Water requirements for dust control and fire suppression
Physical	• Runoff coefficients based on vegetation, soil, geology, and presence of permafrost • Land uses • Water management infrastructure (e.g., ditches, ponds, pipes, culverts, siphons, etc.) • Flow, storage, and pumping capacity
Climate	• Temperature • Precipitation • Evaporation and evapotranspiration • Snowpack • Wind speed and direction • Radiation • Humidity
Hydrological and hydrogeological	• Dewatering volumes and flows • Ice cover • Open water characteristics • Water levels, surface areas, bathymetry, and water volumes in potentially affected water bodies • Groundwater flow rate, flow direction, recharge, rates, and artesian conditions • Mine site drainage
Water quality	• Mass of water quality constituents in mine water management systems • Concentration of constituents in mine effluents • Long-term average concentrations
Geochemical	• Soil, mineral, and rock types • Drainage chemistry

Adapted from Golder Associates (2011).

Table 4.30 Available water and mass balance models (Golder Associates, 2011)

Model	Description
Hydrologic and water quality models	
Generalized Watershed Loading Functions (GWLF)	Typically used to evaluate land management practices and land surface characteristics on downstream point and non-point source sediment and nutrient loading. Does not account for toxins or metals.
Hydrological Simulation Program – FORTRAN (HSPF)	Comprehensive model for watershed hydrology and water quality simulation from point and non-point source loading. Frequently used in Total Maximum Daily Load studies and available for free from the U.S. EPA.
Hydrodynamic models	
Environmental Fluid Dynamics Computer Code (EFDC)	General purpose 3D hydrodynamic numerical model. Can be applied to boundary layer type environmental flows categorized as vertically hydrostatic.
Water Quality Analysis Simulation Program (WASP)	General purpose model for fate and transport studies of conventional and toxic pollutants in surface water. Can be applied to 1D, 2D, and 3D problems and can be linked with other models.
CE-QUAL-W2	A 2D laterally averaged model best applied to stratified water bodies like reservoirs and estuaries. Widely applied to rivers, lakes, reservoirs, and estuaries.
Effluent mixing models	
Cornell Mixing Zone Expert System (CORMIX)	Mixing zone model and decision support system for mixing zones from continuous point source discharges. Used to support EIAs and evaluate discharge compliance. Supported by the U.S. EPA.
Visual Plumes (VP)	Models single and multiple submerged aquatic plumes in stratified ambient flows. Used for marine and freshwater discharges of buoyant and dense plumes.

Excel-Based Deterministic Model

A relatively simple Excel-based deterministic model was developed by Golder Associates (2011) for the Yukon Government and Environment Canada to provide a basic understanding of water quality and water flows for a mining operation across a broad range of operating and climatic conditions. The model utilizes the inputs described in the previous section. This model has three major assumptions (Golder Associates, 2011):

- Sub-watershed ponds are equipped with pumps or discharge structures to evacuate all monthly flows; there is no accumulation of water in the ponds
- Sufficient monthly inflows are available to meet all water demands
- Hydrologic productivity is determined by runoff coefficients, which rely on physical, climate, hydrological, and hydrogeological inputs

The model can be further refined by accounting for site-specific variables and infrastructure such as (Golder Associates, 2011):

- Pumping capacity of all systems
- Storage-elevation curves of ponds and reservoirs
- Discharge-elevation curves of all outlets
- Minimum and maximum operation water levels and physical restraints of retention structures

4.4 ACID MINE DRAINAGE

Acid mine drainage (AMD) is the flow of acidic water from mining operations. Acid mine drainage is produced from the oxidation of sulfide rich minerals, most commonly pyrite (FeS_2), by water and oxygen. The resulting outflow creates acidic conditions that can pollute surface water and groundwater. Furthermore, metals present in the sulfide rich rock may partially dissolve in waters with low pH, increasing toxic metal concentrations and thus the detrimental impact on aquatic life and ecosystems (Jennings et al., 2008).

The chemical and biological reactions associated with AMD are natural processes that usually occur slowly in undisturbed ore, posing little ecological threat (U.S. Environmental Protection Agency (U.S. EPA), 1994). However, in mining activities these processes are greatly accelerated by large land disturbances, greatly increasing contact of sulfide rich minerals to water and oxygen. The reactions responsible for AMD create a cyclical process that perpetuates the continual creation of AMD (Leading Practice Sustainable Development Program for the Mining Industry,

2007b). As a result, acid mine drainage can persist throughout the life of an active mine and long after it has been abandoned. Therefore, any activity that might result in the disturbance of sulfide rich materials should be prefaced with a full evaluation of AMD potential. During this evaluation, emphasis should be placed on preventing and mitigating AMD creation rather than treating AMD (Leading Practice Sustainable Development Program for the Mining Industry, 2007b), as perpetually treating AMD results in high ongoing costs.

Formation of AMD

Acid mine drainage can form from a number of sulfide minerals. Acid mine drainage reactions for pyrite (FeS_2) are perhaps the best known and most studied, as pyrite is the most abundant sulfide mineral and is commonly found in metal and coal deposits throughout the world (Lottermoser, 2010).

Pyrite is initially oxidized by water and oxygen to form sulfate and hydrogen ions (U.S. Environmental Protection Agency (U.S. EPA), 1994):

$$2\ FeS_2(s) + 2\ H_2O + 7\ O_2 \rightarrow 4\ H^+ + 4\ SO_4^{2-} + 2\ Fe^{2+}$$

The ferrous ion (Fe^{2+}) can then be further oxidized to form ferric iron (Fe^{3+}):

$$4\ Fe^{2+} + O_2 + 4\ H^+ \rightarrow 4\ Fe^{3+} + 2\ H_2O$$

For pH values above 4, this reaction can be abiotic, or it can be catalyzed by a variety of naturally occurring sulfide ore bacteria (Johnson and Hallberg, 2005). Below a pH of 4, the abiotic oxidation of the ferrous ion is much slower and biologically mediated oxidation plays an increasingly large role in AMD formation (Johnson and Hallberg, 2005; Mielke et al., 2003). The ferric iron, if in contact with pyrite, can then react to further dissolve pyrite and generate more acid:

$$2\ FeS_2(s) + 14\ Fe^{3+} + 8\ H_2O \rightarrow 16\ H^+ + 2\ SO_4^{2-} + 15\ Fe^{2+}$$

Because ferric iron can act as an oxidizing agent to pyrite, and the oxidization of pyrite produces ferrous iron that is then oxidized back into ferric iron, a cyclic process forms. The regeneration of ferric iron is key to the cyclic process (Johnson and Hallberg, 2005). Biological oxidation of iron plays an important role and it is estimated that bacteria can increase the oxidation of sulfide up to 10^6 times faster than the abiotic rate (Mielke et al., 2003).

The solubility of ferric iron is pH dependent, and thus so is propagation of AMD. Under pH conditions greater than 3, ferric iron can precipitate out of solution as an iron oxide in the form of a yellow, orange, or red deposit in streams (Lottermoser, 2010; U.S. Environmental Protection Agency (U.S. EPA), 1994):

$$Fe^{3+} + 3\ H_2O \rightarrow 3\ H^+ + Fe(OH)_3(s)$$

$$Fe^{3+} + 2\ H_2O \rightarrow 3\ H^+ + FeOOH(s)$$

However, this reaction further contributes to the acidity of the solution, decreasing the pH and allowing more ferric iron to be dissolved, thus, increasing the oxidation of pyrite. As acidity increases, other metals naturally found in the ore, such as zinc, lead, nickel, and copper will readily dissolve, creating further environmental hazards.

Factors Affecting AMD

Complex interactions between geochemical, biological, and hydrologic factors affect AMD potential and formation rates. Waste rock piles and tailings have a large potential for AMD due to their small particle sizes and resulting high surface area. However, both water and oxygen are required for AMD formation, meaning local hydrology and weather can play a key factor in generation rates. Additional factors are summarized below in Table 4.31.

Different sulfide minerals have different potential for AMD due to differing reaction rates and oxidation products. Minerals such as pyrite, marcasite (FeS_2), and pyrrhotite ($Fe_{1-x}S$) are very reactive and oxidation results in highly acidic waters (Lottermoser, 2010). Other minerals such as galena (PbS), millerite (NiS), and covellite (CuS) are less reactive due to the lack of iron released during oxidation, stable crystal structures, and the encapsulation of sulfide minerals by less soluble minerals (Lottermoser, 2010; Plumlee et al., 1999). Other minerals such as cinnabar (HgS) are least reactive and usually do not generate acidic waters (Lottermoser, 2010; Plumlee et al., 1999).

Impacts of AMD

Acid mine drainage primarily affects water quality and aquatic ecosystems (Table 4.32). As AMD runoff enters streams and lakes, it decreases the pH, resulting in major changes in water chemistry and affecting aquatic organisms. pH levels can reach as low as 2.0–4.5 in receiving waters; however,

Table 4.31 Factors affecting AMD potential and generation rates (Jennings et al., 2008; Leading Practice Sustainable Development Program for the Mining Industry, 2007b; Plante et al., 2012; U.S. Environmental Protection Agency (U.S. EPA), 1994)

Factors affecting AMD potential and rates	Comments
Type of sulfide minerals present	Well-crystalized minerals have less surface area; with less area available for oxidation, AMD generation rates decrease. Certain sulfide minerals oxidize easier than others.
Particle size	Smaller particles have larger surface areas and thus, more area available for reactions. Smaller pore sizes may limit oxygen permeation. Waste rock piles and tailings are at high risk for AMD due to their small particle sizes.
Presence of metals	Oxidation of ferrous iron increases acidity. At certain pHs, other metals in the surrounding minerals will solubilize and increase toxicity and acidity.
Amount and type of neutralizing minerals present	Alkalinity produced from reactions with carbonate minerals (e.g., calcite ($CaCO_3$), dolomite ($CaMg(CO_3)_2$), etc.) can neutralize acid production. Some neutralizing minerals are released quicker than others.
Oxygen	Oxygen is necessary in the chemical reaction to oxidize FeS_2. The path of oxidation entry and rate of oxidation should be considered. Materials with large pore sizes have increased oxygen permeability.
Bacteria	Naturally occurring bacteria accelerate reactions, increasing AMD generation rate. These bacteria can survive under low pH conditions.
Water and hydrology	Water is necessary in the chemical reaction to oxidize sulfide ores. If acid generating material is located below the water table, the diffusion of oxygen through water is slowed and retards acid production.
Volume of materials	The larger the volumes of material produced, the greater the volume of waste rock and tailings exposed to water and oxygen.
Exothermic reaction	Sulfide oxidation reactions are exothermic, creating thermal gradients and consequently creating a convective flow that increases oxygen flow around particles, accelerating oxidation.

Table 4.32 Environmental and water quality impacts of acid mine drainage (Jennings et al., 2008; U.S. Environmental Protection Agency (U.S. EPA), 1994)

Type of impact	Effects
Aquatic organisms	Impaired osmotic mechanisms, impaired gill functions, reduced survival, fish kills and death due to hypoxia, overall reduced biodiversity and abundance
Aquatic habitat	Precipitation of iron may coat the stream bed and sediments, reduced availability of clean gravel for spawning, reduced habitat for benthic organisms, smothers aquatic plants
Water quality	Decreased pH, increased sulfate levels, heavy metal contamination, restricted use of receiving waters and groundwater

many fish species are known to be severely impacted beginning at pH of 4.5 to 5.5. Furthermore, dissolved metals such as zinc and copper are highly toxic to many aquatic organisms and acid mine drainage has been known to cause receiving bodies of water to become devoid of life. Heavy metals and decreased pH can also affect downstream communities by contaminating reservoirs and streams, making water resources unfit for irrigation, human consumption, or recreational use without additional treatment. Increased sulfate levels from AMD, although not typically associated with significant human health or ecological impacts can affect the taste of drinking water and may cause corrosion in water infrastructure (Bowell, 2004). Acid mine drainage can also lead to heavy metal and sulfate contamination of groundwater resources due to infiltration.

As acid mine drainage enters a body of water, the pH may rise due to dilution and alkalinity in the receiving waters. If the pH rises high above 3, iron hydroxide will precipitate out of solution. These iron precipitates, known as "yellow boy," can paint a stream yellow–orange and smother the stream bed and benthic habitat.

Prediction and Modeling of AMD

The geology of the surrounding area can provide indicators of potential for AMD and should be analyzed during exploration and well before the construction phase. A detailed investigation needs to follow basic indicator tests and should include mineralogical investigations, elemental analysis, sulfur and carbonate speciation, acid neutralizing capacity, and reactivity analyses.

There are currently several different impact prediction methods, used both individually and in conjunction with one another, including: static tests, kinetic tests, and mathematical modeling. In static tests, the amount of sulfur present in minerals is measured and attributed to being acid forming minerals or non-acid forming minerals. The amount of neutralizing minerals is also measured. The two values of acid forming potential and neutralizing potential are compared in a ratio to predict the probability of acid generation. In essence, a static test predicts the sample's potential to produce acid. Kinetic tests are used to analyze the rate at which acid forms and is neutralized at the mine site. Kinetic tests take longer and are more expensive than static tests but they can more accurately portray the accelerated weathering processes at a mine (U.S. Environmental Protection Agency (U.S. EPA), 1994).

Mathematical modeling of AMD has been performed in an attempt to conserve resources. Due to simplifications employed in mathematical modeling, uncertainties exist and model applicability must be examined on a site-specific basis. The two models used in practice are deterministic models and empirical models.

Deterministic models predict AMD by solving a system of equations that represents the various physical factors interacting during the reaction process (U.S. Environmental Protection Agency (U.S. EPA), 1994). An advantage of deterministic models is that they can potentially be used to predict the changes in AMD processes over a long period of time (U.S. Environmental Protection Agency (U.S. EPA), 1994). A disadvantage of deterministic models is that their use has not been validated or tested extensively in the field (U.S. Environmental Protection Agency (U.S. EPA), 1994).

Empirical modeling involves using extrapolated data from site specific tests and samples to derive equations. As empirical models do not consider the driving mechanisms behind acid mine drainage, model accuracy depends on the quality of testing and sampling. A limitation of empirical models is that they are site specific, so results from one model are not easily transferred to another mine. However, empirical modeling can provide a cost effective estimate of short-term AMD impacts.

Utilizing AMD impact prediction can provide best estimates, but may not always be accurate. In a study by Skousen et al. (2002), a little over 10% of coal mines surveyed did not accurately predict acid generation. Inaccurate predictions can be due to mistakes when characterizing hydrologic and geochemical site factors and/or failure of mitigation techniques. An area of geochemical characterization uncertainty is the ratio of maximum neutralizing potential to maximum acid production potential used in static tests. Skousen et al. (2002) argue that the ratio of neutralization to acid production potential needs to be greater than 2 to avoid AMD while

Jennings et al. (2008) notes that the ratio might need to be greater than 5 to avoid AMD in the long term. Due to the various mistakes that can be made during characterization, prediction techniques are not routine or robust and should be considered an area of uncertainty and on-going research (Jennings et al., 2008).

4.5 LAND IMPACTS FROM MINING

Mining operations have significant impact on the physical characteristics of the land (Table 4.33). Land impacts, combined with other environmental impacts, affect the land use patterns of the local area. In turn, land use patterns have ecological, economic, social implications.

In 1977, the U.S. Surface Mining Control and Reclamation Act was passed to regulate and enforce the reclamation of surface mines and minimize land impacts, specifically for coal mining. Coal mining operators are required to pay into the Abandoned Mine Reclamation Fund, which is distributed to fund mining reclamation and other activities such as control of erosion, landslides, and land subsidence.

Table 4.33 Land attributes affected by mining and mineral processing

Land attributes	Effect on health or environment	Source
Soil contamination	Reduced vegetative and animal growth. Contaminants can enter the food chain	Erosion, particulate matter, waste disposal, mine water, vehicle traffic
Topsoil disturbance	Reduced vegetative growth and biodiversity	Land clearing, erosion
Erosion	Increases TSS and TDS in streams, leading to decreased use of water. Increases PM in air, leading to health problems	Wind and water erosion increases in areas with less vegetation and cover
Subsidence	Decreases aquifer capacity, diverts surface and ground water resources, damages infrastructure, creates safety hazards	Underground mine design, geological properties
Land use patterns	Loss of agricultural land and forests, decrease in value of adjacent land parcels, population movements	Conflicting types and intensities of land use, impacts to local/regional air, water, ecosystem and soil

Soil Contamination

Soil contamination is an ongoing problem that affects active mines and can persist long after mine remediation. Mining involves the excavation of massive amounts of buried materials and bringing them to the surface for processing. The introduction of materials that contain heavy metals and other contaminants not normally found in surface soils creates the potential for soil contamination (Environmental Law Alliance Worldwide, 2010). Processing raw ore requires the use of potentially hazardous chemicals and creates waste products such as tailings, waste rock, slag, and muds. These waste products are usually stored in open air piles or tailings ponds. The failure or leakage of tailings ponds and the erosion or mismanagement of waste piles can contaminate large areas of land if not properly managed (U.S. Environmental Protection Agency (U.S. EPA), 1997a). Atmospheric emissions, primarily particulate matter, from all stages of mineral processing and from wind erosion can contain contaminants such as heavy metals, inorganic ions, and other compounds. Once in the atmosphere, these compounds can then contaminate the soil via atmospheric deposition. Contaminants may also be directly spread throughout a mine site by vehicular traffic during loading, unloading, and transportation activities. Mine water, AMD, and waste chemicals, if not adequately managed, can also act as a transport mechanism for the contamination of soils near mining operations.

Once in the soil, hazardous compounds, such as copper or lead, can be absorbed by plants or ingested by animals, negatively impacting health and entering the food chain in the process. Once in the food chain, compounds such as mercury, copper, or lead can bioaccumulate in higher trophic level species. Soil contamination can directly impact agricultural crops and negatively impact the lifestyles of aboriginal peoples living off of the land. Furthermore, if not managed correctly, contaminated soils can enter nearby streams or lakes during erosion or precipitation events, reducing water quality and impacting aquatic species. Soils contaminated from hazardous chemicals may also pose a direct contact risk if used as fill material or for soil supplements (U.S. Environmental Protection Agency (U.S. EPA), 1997a).

Heavy Metal Contamination

One of the greatest concerns of mining and mineral processing is heavy metal contamination of the soils in and around the operation. Heavy metals are naturally found in most soils are low concentrations. However, the excavation of metal rich ore and waste rock to the surface provides a contamination pathway. Heavy metal contamination is not limited to metal mining; metal containing ore and waste rock are often excavated as a by-product in other mineral mining operations. Heavy metal contamination from mining can occur due to atmospheric deposition from combustion exhaust or wind erosion, crushing, milling, and the improper disposal of high metal wastes including tailings and wastewater. Coal ash and coal slag can also be sources of heavy metal contamination. Historically, mining has not always been strictly regulated and has a legacy of widespread heavy metal contamination in soils throughout the world (Wuana and Okieimen, 2011). Wastewaters from such operations have been historically applied to land, causing the accumulation of heavy metals in soils over long periods of time. Extensive mining and smelting of heavy metals, such as lead and zinc, and the improper disposal of tailings and wastes have left contaminated soils that pose a risk to human and environmental health. Additional details on the atmospheric and water impacts, as well as human and ecological health impacts, from heavy metals such as mercury and lead can be found in Sections 4.2 and 4.3.

A study of 72 mining areas across 22 provinces in China by Li et al. (2014) found that soil heavy metal contamination is prevalent throughout industrial and mining areas in southern China. In the study, chromium and nickel were found to be of little concern, as 0% and 12.6% of sites were considered moderately to heavily contaminated for chromium and nickel, respectively. Copper, lead, zinc, and cadmium, on the other hand, had the highest contamination levels, with 31.4%, 51.5%, 32.4%, and 71.2% of sites being moderately to heavily contaminated, respectively. Concentrations of heavy metals in soil at select mining sites around the world are presented in Table 4.34.

The fate, transport, bioavailability, and toxicity of many heavy metals are related to their chemical form and speciation (Table 4.35). For example, many heavy metals exhibit high mobility under acidic conditions. Heavy metals introduced into the soil are initially redistributed into various chemicals forms by fast and slow adsorption reactions including (Wuana and Okieimen, 2011):

- Mineral precipitation and dissolution
- Aqueous complexation
- Biological immobilization or mobilization
- Plant uptake
- Ion exchange
- Adsorption or desorption

Table 4.34 Mean heavy metal concentrations in soil at mining sites throughout the world

Location	Number of mines examined	As	Cd	Cr	Cu	Ni	Pb	Zn	Hg
China	72	195.5	11.0	84.28	211.9	106.6	641.3	1163	3.82
Iran	3	146.2	1.49	–	88.4	–	1002	363.4	3.13
Spain	16	191.9	6.59	63.2	120.8	28.35	881.8	465.8	52.9
South Korea	70	70.8	1.99	–	79.09	22.00	111.1	183.2	1.12
Vietnam	3	3144	135	1501	271.4	2254	30,635	41,094	–
India	5	18.62	3.82	1509	63.49	1069	304.7	338.8	–
U.S. EPA heavy metal cleanup threshold	–	25	70	230	–	1600	400	23,600	–

All values are in mg/kg.
Adapted from Li et al. (2014), U.S. Department of Agriculture (2000), U.S. Environmental Protection Agency (U.S. EPA) (2002).

Table 4.35 Health impacts due to heavy metal exposure (Agency for Toxic Substances and Disease Registry, 2005; Eisler, 1985a,b, 1986, 1988, 1993, 1998a,b; Govind and Madhuri, 2014; Martin and Griswold, 2009; Solomon, 2008; U.S. Department of Agriculture, 2000; Wuana and Okieimen, 2011)

Heavy metal	Human health effects	Aquatic effects	Terrestrial effects	Bioaccumulation and exposure risk
Cadmium	Negative impacts on kidneys, liver, and gastrointestinal tract, lung damage, fragile bones, increased risk of cancer	Fish skeletal deformities and impaired kidney functions. Decreased aquatic plant growth. Reduced survival of invertebrates. Generally more toxic in freshwater than saltwater ecosystems.	Reduced plant and root growth.	Accumulates in fish, plants, and animals. Exposure from roots, rice, liver, meat, shellfish, mushrooms, and seaweed.
Arsenic	Skin poisoning, nausea, vomiting, affects kidneys, circulatory system, and nervous system, increased risk of cancer, death	Reduced algal, fish, and amphibian growth, reduced aquatic biomass, reduced reproductive success, impaired fish behavior, increased invertebrate mortality.	Weakness, vomiting, diarrhea, and increased mortality and morbidity in mammals. Slowness, incoordination, and falling in birds.	Toxic to plants at levels not harmful to humans. Usually not absorbed in plants or limited to roots. Accumulates in fish and shellfish. No evidence of biomagnification.
Chromium	Allergic dermatitis, respiratory irritation, negative impacts on kidneys, liver, and nervous system, increased risk of cancer	Inhibits fish and algal growth, increased vulnerability to predation.	Carcinogenic in mammals and birds.	High uptake potential in bottom feeders and possible uptake in plants. Possible exposure from roots.
Zinc	Trace nutrient; can cause kidney damage and gastrointestinal irritation	Hyperactivity, sluggishness, gill damage, and death in fish.	Muscular weakness and inability to swim in birds. Inhibition of plant and root growth. Joint damage in mammals.	Accumulates in fish and some plants.

Continued

Table 4.35 Health impacts due to heavy metal exposure (Agency for Toxic Substances and Disease Registry, 2005; Eisler, 1985a,b, 1986, 1988, 1993, 1998a,b; Govind and Madhuri, 2014; Martin and Griswold, 2009; Solomon, 2008; U.S. Department of Agriculture, 2000; Wuana and Okieimen, 2011)—cont'd

Heavy metal	Human health effects	Aquatic effects	Terrestrial effects	Bioaccumulation and exposure risk
Copper	Trace nutrient; can cause anemia, liver and kidney damage, and gastrointestinal irritation	Gill damage, interferes with osmoregulation and metabolism in fish. Increases membrane permeability in algae. Inhibited photosynthesis in plants.	Inhibits plant and root growth. Reduced animal growth and possible liver, kidney, brain, and muscle damage.	Usually does not accumulate in vertebrates. Accumulates in mollusks, crustaceans, and macrophytes.
Nickel	Trace nutrient; can cause allergic dermatitis, allergic reactions, and gastrointestinal irritation	Convulsions, loss of equilibrium, gill damage, and death in fish. Inhibits algal, fish, and crustacean growth.	Inhibits growth and reduces bone densities in birds. Can inhibit growth in tolerant plants.	Not known to accumulate in animals. May accumulate in some plants.
Selenium	Nausea, selenosis, hair loss, neurological abnormalities, and respiratory irritation	Reproduction failure and death in fish. Reduced growth in algae and fish.	Abnormal growth and deformities in birds. Infertility and high mortality in mammals.	Readily bioaccumulates in plants, fish, crab, and lobster. Some plants require high selenium levels and can be potentially harmful to humans.
Mercury	See Table 4.14	See Table 4.14	See Table 4.14	Accumulates in fish, shellfish, and animals eating fish. Typically accumulates in plants in a non-bioavailable form to humans and animals.
Lead	See Table 4.19	See Table 4.19	See Table 4.19	Usually not absorbed in plants or limited to roots. Exposure due to root plant tissues, soil, or dust adhering to plants.

Heavy metals contamination is of significant concern because unlike organic contaminants, they usually cannot undergo microbial or chemical degradation (Wuana and Okieimen, 2011). Mercury and selenium are notable exceptions, and can be transformed and volatilized by microbes (U.S. Department of Agriculture, 2000). Once the soil becomes contaminated, heavy metals will persist unless the soil is physically removed and treated or the metals are removed through leaching, volatilization or phytoremediation. Acid mine drainage, in particular, is of great concern for soils contaminated with heavy metals, as the decrease in soil pH can solubilize many heavy metals and transport them to contaminate nearby surface water and groundwater resources.

Exposure to heavy metals is typically chronic, via ingestion through the food chain, although acute heavy metal poisoning is possible (U.S. Department of Agriculture, 2000). The most common heavy metal exposure pathways are the soil-plant-human and soil-plant-animal-human pathways. Plants generally do not uptake or accumulate significant amounts of heavy metals in fruiting crops, thus, they may be safe to eat. However, heavy metal contamination is a high risk for leafy vegetables and root crops, as plants tend to accumulate heavy metals in their leaves and root crops are in direct contact with the soil (U.S. Department of Agriculture, 2000; Wuana and Okieimen, 2011). All heavy metals pose a risk to small children who may directly ingest contaminated soil, either by accident or on purpose (pica) (Wuana and Okieimen, 2011).

Although some metals are necessary as trace nutrients for proper growth, excessive exposure is known to negatively impact plant and animal life. In plants, heavy metals are known to cause oxidative stress, negatively affecting photosynthesis, chlorophyll fluorescence, stomatal resistance, reproductive processes, and ultimately reduced growth or death (Abdul-Wahab and Marikar, 2012). Bioaccumulation presents the greatest risk in animals, as it leads to chronic heavy metal exposure and reduces animal fitness, interferes with reproduction, and can lead to cancer or death. In animals, heavy metals are known to affect the predator-prey relationship by reducing the prey's ability to detect and respond to predators (Boyd, 2010).

Topsoil Disturbance

The soil that covers the earth provides nutrients and the means for sustaining life, both directly and indirectly (Jain et al., 2012). Topsoil, the top several inches of soil, provides nutrients and growing space for plant life, supports beneficial microorganisms, a supply of seeds and spores, and promotes the rapid development of groundcover (Leading Practice Sustainable Development Program for the Mining Industry, 2006). Nutrients and microbial communities found in the topsoil essential to plant growth are not typically found deeper in the soil profile and must be preserved (Leading Practice Sustainable Development Program for the Mining Industry, 2006).

Topsoil quality and quantity can be altered through a range of forces, including human activities (mining) and erosion. During surface mining operations, the clearing of topsoil and all the plants and trees that rely on it is inevitable. This complete loss of habitat and ecosystem disruption can devastate large areas of natural landscape, much of which was originally natural areas such as forests, wetlands, or grasslands. This translates to a direct loss of habitat for local flora and fauna, contributing to habitat fragmentation and intrusion of weed species and feral animals (Environment Australia, 2002). Reductions in habitat affect creatures that live and forage in those areas, reducing overall carrying capacity and increasing the likelihood of extinction events. Adjacent habitats are often impacted as well, as land must be cleared for the development of camps, towns, and infrastructure vital to the success of a mining project (Environment Australia, 2002). Land clearing of this nature can lead to the pollution and compaction of vital topsoil as well as dilution due to mixing with lower layers of overburden and the loss of vital beneficial microbial communities, hindering rehabilitation efforts (Environment Australia, 2002; Mummey et al., 2002).

After topsoil is removed and stockpiled for mining operations, complete restoration to pre-mining conditions is unfeasible. Reclaimed surface mining sites have shown that while plant cover may not be significantly different between reclaimed and undisturbed sites, plant species diversity was significantly lower in reclaimed areas (Mummey et al., 2002). Soil microbial communities in a reclaimed surface mining site exhibited microbial biomass levels of 14–44% when compared to nearby undisturbed soils, even after a 20 year period, and levels of beneficial arbuscular mycorrhizal fungi (which form symbiotic relationships with plants, encouraging nutrient uptake and growth) have been shown to be greatly decreased by soil storage and disturbances associated with mining (Harris et al., 1987; Mummey et al., 2002; Stahl et al., 1988). Furthermore, availability of ammonia and nitrate in reclaimed soils was found to be significantly lower than in undisturbed soil, indicating inefficient nitrogen cycling and lower likelihood of nitrogen uptake by plants and microbes

(Mummey et al., 2002). While complete restoration is unfeasible, mining remediation operations will aim to restore the land to a condition suitable for agreed upon land uses.

Erosion

Erosion can be caused by wind forces or hydraulic forces, and is often exacerbated by mining activities. Wind erosion can result in air quality problems through dust creation. Hydraulic forces, such as rainwater or drainage, can result in high sediment loads in surface waters, negatively affecting aquatic species. Erosion occurs on exposed land areas, including but not limited to stockpiles, waste dumps, and exposed tailings. Wind erosion can be reduced through the mitigation measures including wind fences, chemical binders, and increasing the moisture content of the soil (Leading Practice Sustainable Development Program for the Mining Industry, 2009). Hydraulic erosion can be reduced through stormwater erosion controls implemented in early stages of a mine's life. In addition to stormwater erosion controls, erosion can be managed through mine planning. Open areas should be minimized and vegetation cleared only when necessary. Early rehabilitation and re-vegetation should be performed on open areas that are no longer active. Erosion of stockpiles and waste dumps can be managed through the establishment of ground cover and vegetating surfaces (Environment Australia, 2002).

Erosion Modeling

Designing mine features to limit land impacts requires the comparison of different alternatives. To reduce soil disturbance, and subsequent effects on air and surface water quality, the soil erosion potential of alternative designs can be predicted through several different types of models, including empirical and physical models.

Empirical models are based on observed effects from environmental factors that are related to erosion by statistical analysis. An example of an empirical based model is the Universal Soil Loss Equation. This empirical model takes into account a rainfall erosivity factor (R), a soil erodibility factor (K), topographic features including slope and length (S and L), vegetative cover (C), and erosion control practice factor (P). The product of these factors produces the computed soil loss for a given storm in tons/ha (A).

$$A = R * K * (LS) * C * P$$

Tables of values and equations for calculating variables can be found in *Groundwater Hydrology* 3rd ed. (Todd and Mays, 2005).

Process (or physical) models represent the processes involved in erosion, including detachment, transport, and sedimentation. Computer simulations can take into account hydrology, hydraulics, geology, and erosion mechanics. The Water Erosion Prediction Project (WEPP) is software used by the USDA and U.S. EPA that predicts erosion potential based on (Flanagan et al., 2007):

- Climate (rainfall, temperature, solar radiation, wind)
- Winter (freeze-thaw cycles, snow accumulation, snow melting)
- Irrigation (stationary sprinkler, furrows)
- Hydrology (infiltration, depressional storage, runoff)
- Water balance (evapotranspiration, percolation, drainage)
- Soils (types and properties)
- Land use (cropland, rangeland, forestland)
- Residue management and decomposition
- Tillage impacts on infiltration and erodibility
- Deposition (rills, channels, impoundments)
- Sediment delivery, particle sorting and enrichment

WEPP can be applied to a single hillside or more complex watersheds (Flanagan et al., 2007).

Wind erosion was most commonly modeled using the empirical Wind Erosion Equation (WEQ), originally developed in for agricultural fields (Hagen, 2010):

$$E = f(I, K, C, L, V)$$

where

E = long-term annual soil loss per unit area

I = soil erodibility

K = soil ridge roughness

C = climatic factor

L = field length

V = vegetative factor

Shortcomings in the WEQ, including failures to predict erosion where it was observed led to the development of the complex Wind Erosion Prediction System (WEPS) by the U.S. EPA, Bureau of Land Management, and the National Resource Conservation Service to replace the WEQ. The WEPS models erosion on an sub-hourly basis by simulating daily weather and soil conditions, common management practices, wind erosion processes, vegetative cover, and hydrology (Hagen, 2010).

Subsidence

Mining subsidence is the lowering or collapse of the land surface, typically caused by underground mining activity. Subsidence occurs when rocks above the mine cannot provide adequate support and collapse from the weight of the overlying material. This can occur when a mine is active or long after a mine has been abandoned (Ge et al., 2007). Whether or not subsidence will occur is affected by several different parameters including (Ge et al., 2007):

• Depth of cover
• Overlying strata properties
• Seam thickness
• Panel width
• Pillar size
• Surface topography

There are two main types of underground coal mining: longwall mining and room-and-pillar mines. Longwall mining is when an entire seam of minerals is removed from the ground and the rocks above are allowed to collapse, resulting in planned subsidence. Longwall mines are typically designed to result in small amounts of subsidence at the land surface; usually less than the width of the coal panel extracted (Mine Subsidence Engineering Consultants, 2007). Longwall mines can be operated in such a way to avoid subsidence; however, this involves mining panels relatively small in width, which does not maximize coal utilization or economic returns (Mine Subsidence Engineering Consultants, 2007).

A room-and-pillar mine is when underground mineral seams are mined in sections and columns are left in place to keep the roof from collapsing. These pillars, which can sometimes account for over half of the recoverable coal, are not always strong enough to support the weight of overlying material and may result in the formation of troughs or sinkholes. Sinkholes typically occur in mines that have less than 50 feet between the surface and the mined area (Pennsylvania Department of Environmental Protection, 2012). These usually occur in older, abandoned mines as most current mines are much deeper than 50 feet. Troughs can occur in both active and abandoned mines due to the punching of pillars through a soft floor or roof, or the sagging of a roof material (Pennsylvania Department of Environmental Protection, 2012).

Another way that subsidence can occur at an abandoned or rehabilitated mine is through the reinstatement of a water course over the mined backfill (Leading Practice Sustainable Development Program for the Mining Industry, 2008) or through the ponding of water on top of a landform (Leading Practice Sustainable Development Program for the Mining Industry, 2006).

Environmental Effects of Subsidence

Subsidence can cause structural damage to overlying buildings and infrastructure, including railway tracks, roads, and homes, threatening human safety. In some areas, mineral rights are separate from property ownership, meaning mining companies are allowed to recover mineral resources from beneath homes and other buildings. In other cases, mining operations have ceased long ago and expanding suburbs have encroached above abandoned mines. Some types of subsidence may be more dangerous than others, but they all require costly infrastructure improvements.

In addition to structural damage, subsidence can change natural drainage patterns, affecting water resources and the environment (Bell et al., 2000). Subsidence can result in the formation of sinkholes, surface depressions, or cracks, all of which can divert water from nearby lakes or steams into the underlying strata, resulting in a partial or complete water loss (Pennsylvania Department of Environmental Protection, 2012). Not only can this result in complete habitat destruction for aquatic species, it can also impact surface water resources used for drinking water or recreation. Groundwater resources can be affected in a similar manner, where groundwater is directly drained through subsidence induced cracks into the underground mine, resulting in wells and springs running dry (Pennsylvania Department of Environmental Protection, 2012).

Smaller subsidence events can also impact drainage in canals, sewers, agricultural fields, and natural stream systems. These impacts can be significant enough to decrease bank stability, whereby increasing erosion rates and sedimentation, negatively impacting water quality (Hughes, 2005). Small changes in hydrological regimes can also have major impacts on wetland ecosystems by changing the quantity and flow patterns of water, affecting sedimentation, erosion, and the distribution of plant and animal species (Hughes, 2005).

Land Use Patterns

Land use can be generally defined as the benefits or products from the management or use of land. Various land uses include grazing, forestry, recreation, traditional indigenous uses, conservation, water catchment, agriculture, and rural and urban development. Changes in land use patterns often have secondary effects on air and water quality, ecology, and the economy (Jain et al., 2012). Schueler et al. (2011) noted that mining results in environmental and ecosystem degradation that decreases land use purposes. Surface mining, for example, removes vegetation and soils, interrupts ecosystem service flows, and results in inevitable and often permanent deforestation and loss of farmlands and natural habitats. Mining activities also frequently produce toxic waste that causes water pollution and health problems (Leading Practice Sustainable Development Program for the Mining Industry, 2008), making it difficult for indigenous communities to maintain their traditional lifestyles or for farmers to irrigate their crops. Similarly, dust pollution from heavy traffic on mining dirt roads affects neighboring communities (Leading Practice Sustainable Development Program for the Mining Industry, 2009), and soil erosion is common around mines, leading to barren landscapes with reduced land uses (Leading Practice Sustainable Development Program for the Mining Industry, 2009).

Typically, mining in the developing world often forces populations to relocate and farmers to develop alternative income strategies (Schueler et al., 2011). As a consequence, conflicts between communities and mining operators over land use rights are common in many regions worldwide (Hilson, 2002; Schueler et al., 2011). Some of these conflicts may be resolved by paying compensation or royalties to affected parties; however, conflicts can become especially heated when mineral deposits are located on lands that are considered sacred to native communities, as no amount of money can make up for the cultural and spiritual loss of these areas (Yirenkyi, 2008). In these cases, public outreach programs and continual cooperation with local communities during the early stages of development are necessary to resolve land use issues.

In the past, many mining projects throughout the world had little regard for environmental impacts, to the point where some abandoned sites have no land use possibilities and are still undergoing remediation to this day. Communities, governments, and stakeholders are increasingly demanding that mine operators rehabilitate the land and minimize environmental damage to the point where it is suitable for beneficial use after mining operations cease (Environment Australia, 2002). Industry has responded by conducting environmental impact assessments and implementing environmental management systems and rehabilitation programs to minimize impacts. The incorporation of buffer zones between a mining site and surrounding properties can minimize incompatibilities between adjacent areas with different land uses. This response has improved public image and community relations, as mining (in some cases) becomes a temporary land use that can be integrated with some current and future land uses (Environment Australia, 2002).

4.6 ECOLOGICAL IMPACTS FROM MINING

Jain et al. (2012) noted that ecology is the study of the interrelationships among organisms (including people) and their environment. In the context of this discussion, ecology is used to include those considerations covering plant and animal species. Table 4.36 briefly summarizes the ecological attributes affected by mining.

It is generally agreed that increased biodiversity positively affects ecosystem services, such as provisioning, regulating, and cultural services. Thus, one approach to measuring the degradation of ecological systems is by noting decreases in wildlife species and populations. Since many types of outdoor activities are based directly on wildlife species, there may be an economic, as well as a moral and aesthetic basis for maintaining large, healthy wildlife populations. Hunting and fishing activities provide the

Table 4.36 Ecological attributes affected by mining activities (International Council on Mining & Metals (ICMM), 2006)

Ecological attributes	Effect on health or environment	Source
Terrestrial vegetation and fauna	Reduction of diversity	Land clearing, extraction, blasting, digging/hauling, air pollution, building infrastructure, and population growth
Aquatic fauna	Reduction of diversity	Waste dumping/disposal, processing/chemical use, tailings management, effluent discharges, building infrastructure, and water supply
Aquatic, riparian, and groundwater dependent vegetation	Reduction of diversity	Waste dumping/disposal, processing/chemical use, tailings management, effluent discharges, building infrastructure, and water supply

difference between existence and relative affluence for many persons engaged in services connected with these outdoor recreational pursuits, and in some areas of the country provide an essential component of the human diet. If fisheries and wildlife populations utilized by indigenous groups are potentially affected, the need to proceed carefully with mining projects becomes even more critical (Jain et al., 2012).

In considering the impact of human activities on the biota, it can be determined that there are at least three separable types of interests. The first, *species diversity,* includes the examination of all types and populations of plants and animals, whether or not they have been determined to have economic importance or any other special values. The second area, *system stability,* is concerned with the dynamics of relationships among the various organisms within a community. A third important area, *managed species,* deals with agricultural species and those non-domesticated species known to have some recreational or economic value. Wild managed species are usually overseen by state or other conservation departments under the category "wildlife management." Agricultural species have economic and cultural value and due to their close ties to human needs, may result in controversy if mining activities affect agriculture, especially if the quality or safety of the human food supply appears threatened (Jain et al., 2012).

All the areas in ecology are very difficult to quantify, often being almost impossible to present in terms familiar to scientists of other disciplines. Furthermore, there are literally millions of possible pathways in which interactions among plants, animals, and the environment may proceed. To date, even those scientists knowledgeable in the field have been able to trace and analyze only a small minority of these, although thousands more may be inferred from existing data. Thus, many impacts predicted cannot be absolutely verified. Other interactions are probably appropriate by comparison with known cases involving similar situations, while many more are simply predicted on the basis of knowledge and experience in a broad range of analogous, although not entirely similar, systems (Jain et al., 2012).

The question of chance effects is also an important one in ecology. One may be able to say that the likelihood of significant adverse impact following a certain activity is low, based on available experience. This is definitely not the same as saying that the impact, if it develops, is not serious. The impact may be catastrophic, at least on a regional basis, once it develops. When one works with living organisms, the possibility of the spread from an area where little chance of damage exists to one in which a greater opportunity for harm is present is itself a very real danger. The vectors of such movement cannot be predicted with any accuracy; however, the basic principles best kept in mind are simple enough. Any decrease in species diversity tends to also decrease the stability of the ecosystem, and any decrease in stability increases the danger of fluctuations in populations of economically important species (Jain et al., 2012).

Many other scientific disciplines are often closely related to ecology. When the question of turbidity of water in a stream is examined, for example, it will be found that this effect not only is displeasing to the human observer but has ecological consequences also. Excessive turbidity may cause eggs of many species of fish to fail to develop normally and can render the water unsuitable for the very existence of several species of fish. The smaller animals and the plant life once characteristic of that watershed may also disappear. Thus, the turbidity of the water, possibly caused by land-clearing operations in the stream watershed, may have effects ramifying far beyond the original, observed ones. Similarly, almost all effects relating to the quality of water will also have ecological implications, in addition to those already of interest from a water resources viewpoint. The quantity of water was once a relatively minor concern, but the consequences of an action resulting in significantly less (or more) water may be extremely important. Projects that modify stream flow may also have effects on wildlife and agriculture in downstream areas (Jain et al., 2012).

Other impacts may be simply aesthetic in nature, damaging the appearance of a favorite view, for example. They may also be symptoms of effects that could possibly be harmful to humans if ignored, such as heavy metal accumulation in birds and fish. If we are to view the area of biology, or ecology, in perspective, we must realize that it includes a wide variety of messages to us. These should be interpreted as skillfully as possible, if our future is to be assured (Jain et al., 2012).

Mining Specific Impacts on Ecology

Ecological impacts specific to mining include deforestation, destruction of vegetation and habitats, and reduction of air, water, and land quality, all of which affect the existence of large and small animals, birds, fish, invertebrates, and plants (Table 4.37). Separating ecological impacts from air, water, and land impacts is often difficult as all three often result in significant ecological impacts as well. Specific details on ecological impacts related to reductions in air, water, and land quality from mining and mineral processing can be found in previous sections. Habitat loss,

Table 4.37 Mining activities that can negatively impacts habitats, biodiversity, and regional flora and fauna

Mine activity	Environmental effect	Examples of biodiversity impact
Extraction	Land clearing	Loss of habitat, introduction of plant disease, siltation of watercourses
Blasting	Dust, noise, vibration	Smothering stomata, disturbance of fauna
Digging and hauling	Dust, noise, vibration, water pollution	Disruption of watercourses, impacts on aquatic ecosystems due to changes in hydrology and water quality
Waste dumping	Land clearing, water and soil pollution	Loss of habitat, soil and water contamination sedimentation, acid mine drainage
Processing/Chemical use	Toxicity	Loss of species (fish kills, for example) or reproductive impacts
Tailings management	Land clearing, water pollution	Loss of habitat, toxicity, sedimentation, water quality and stream flow
Air emissions	Air pollution	Loss of habitat or species
Effluent discharges	Water pollution	Loss of habitat or species, reduced water quality
Building workshops and other structures	Land clearing, soil and water pollution	Loss of habitat, contamination from fuel, waste disposal
Waste disposal	Oil and water pollution	Encouragement of pests, disease transfer, contamination of groundwater and soil
Building power lines	Land clearing	Loss or fragmentation of habitat
Provision of accommodation	Land clearing, soil and water pollution, waste generation	Loss of habitat, sewage disposal and disease impacts, pets, disturbance of wildlife
Building roads and railways	Land clearing	Habitat loss or fragmentation, waterlogging upslope and drainage shadows down slope
Population growth	Land clearing or increased hunting	Loss of habitat or species, stress on local and regional resources, pest introduction, clearing
Water supply (potable or industrial)	Water abstraction or mine dewatering	Loss or changes in habitat or species composition

Adapted from International Council on Mining & Metals (ICMM) (2006).

fragmentation, and pollution from mining activities have the most profound impact on local and regional ecology.

Habitat Loss and Fragmentation

Habitat loss and fragmentation often occur simultaneously and represent the greatest threat to many species around the world. Land clearing and alterations from mining activities can destroy native habitats, devastating local fauna and flora and creating conditions that are more favorable for pest species. Habitat fragmentation can be widely defined as the decrease and division of natural habitats and landscapes and includes:

- Fragmentation of one habitat area into many smaller areas
- Reduction of total habitat area
- Isolation of habitat areas from others
- Decreased size of fragmented habitat areas
- Increased habitat edge ratio

Habitat loss and fragmentation can result from natural causes, such as fires, floods, volcanic activities, and landslides. However, habitat loss and fragmentation are more commonly associated with development activities. Various mining and mineral processing activities, including facility construction, land clearing, road building, and building the infrastructure required for producing and transporting product from the mine can contribute to both habitat loss and fragmentation. The impacts of mining operations are particularly significant due to the remote locations of some mineral deposits and the likelihood that development will be in a previously undeveloped area. Mining roads and railways are of primary concern in terms of habitat fragmentation, as they span long distances, form a vast interconnected network, and the constant traffic and maintenance can contribute to the irreversible fragmentation of habitats throughout the world. Habitat loss and fragmentation are not only limited to terrestrial habitats; aquatic habitats may be destroyed or become fragmented due to surface water diversion, overutilization, water pollution from sources such as acid mine drainage, and the building of dams. Greenpeace (2010) has estimated that in the U.S., coal mining alone has altered or destroyed 2.4 million hectares of natural landscape and habitat from 1930 to 2000. An additional 12,000 miles of rivers and streams and 180,000 acres of lakes and reservoirs in the U.S. are estimated to be polluted from mine effluent (Madden and Camcigil, 2000). In China, an estimated 3.2 million hectares of land and natural habitat were destroyed due to coal mining as of

2004 with an additional 47,000 hectares destroyed annually (Greenpeace International, 2010; Yang, 2007).

The results of many ongoing studies have shown that reduced habitat area decreases local flora and fauna populations within habitat fragments and also strongly reduces species richness, sometimes altering the composition of entire communities (Haddad et al., 2015). Decreased habitat area reduces available ecosystem services and the ability for the land to support large populations. Mobile species such as game animals, birds, and predators may be able to move elsewhere, but more sedentary animals, such as burrowing animals, small mammals, and many invertebrates will be greatly impacted by habitat loss (Environmental Law Alliance Worldwide, 2010; U.S. Environmental Protection Agency (U.S. EPA), 1997a). Increases in habitat edge ratios can shift the physical environment, making conditions more favorable for select species; altering predation, herbivory, seed pools, and light penetration may favor pioneer trees, leading to a loss of plant biodiversity that can impact animal species as well (Haddad et al., 2015). Changes in biodiversity and local ecology can create conditions that favor pest species over indigenous species. Isolation decreases the movement of species between isolated fragments and can have genetic consequences such as increased occurrences of inbreeding, which in turn increases the likelihood for local extinction events (Environmental Law Alliance Worldwide, 2010; Haddad et al., 2015). Increased edge ratio, isolation, and reduced habitat areas degrades productivity, pollination, and carbon and nitrogen retention within the fragments (Haddad et al., 2015).

The long-term effects of habitat fragmentation include extinction debt, immigration lag, and ecosystem function debt. Experiments have shown temporal lag in the decreases in species richness (i.e., extinction debt) after fragmentation. Decreases in species richness ranging from 20–75% have been observed, with an average loss of >20% after 1 year, and a loss of >50% after 10 years. These effects occur much slower in larger fragments, and ongoing extinction debt has been observed for over 20 years in the longest experiments (Haddad et al., 2015). Immigration lag, or the slowing of species accumulation in habitat fragments was also observed, with 5% and 15% fewer species in small or isolated fragments after 1 year and 10 years, respectively. Overall, these effects also contribute to ecosystem function debt, whereby fragments show an alteration in nitrogen and carbon soil dynamics, biodiversity loss, changes to plant and animal biomass, reduction in decomposition, simplification of food webs, and overall changes in biotic and abiotic conditions (Haddad et al., 2015). These ecosystem function losses can be as great as 30% after 1 year, and up to 80% after 10 years for small isolated fragments when compared to larger connected fragments (Haddad et al., 2015).

Noise Pollution

Noise pollution can be defined as any disturbing or unwanted noise that interferes or harms humans or wildlife. Although noise constantly surrounds us, noise pollution generally receives less attention than water quality and air quality issues because it cannot be seen, tasted, or smelled. Noise generated by mining operations is often of higher intensity than natural noise, and mining operations can occur throughout the night. Common mining and mineral processing activities that contribute to noise pollution include overburden removal, drilling and blasting, excavating, crushing, loading and unloading, vehicular traffic, and the use of generators.

Noise pollution has a negative impact on wildlife species by reducing habitat quality, increasing stress levels, and masking other sounds. Chronic noise exposure is especially disruptive for species that rely on sound for communication or hunting (Bayne et al., 2008). Animals that use noise for hunting, such as bats and owls, and prey species that rely on noise to detect predators may have decreased patterns of foraging, reducing growth and survivability (Barber et al., 2010; Kight and Swaddle, 2011). Additionally, bird species that rely on vocal communication and other various species, such as nocturnal animals, haven been shown to avoid areas with noise pollution (Barber et al., 2010; Bayne et al., 2008). Reductions in bird populations and foraging activities can in turn negatively impact seed dispersion, affecting ecosystem services and diversity (Francis et al., 2012). Because much of the noise pollution in natural habitats is caused by vehicle traffic, generators, and development in general, noise pollution often exacerbates the problems associated with habitat destruction and fragmentation (Barber et al., 2010).

4.7 ECONOMIC IMPACTS FROM MINING

Mining has both direct and indirect impacts on the local and regional economy in which a mine is located. Economic impacts are somewhat ambiguous and difficult to measure in general, although direct impacts are more easily measured. The contributions of mining and mineral processing to the socio-economic well-being of surrounding regions are still poorly understood, primarily due to a lack of studies (Reeson et al., 2012). Most of the available literature is not from peer reviewed sources or comes from the mining industry, where there is an inherent conflict of interest (Deller and Schreiber, 2013).

There are many different linkages between the mining industry and impacts on the economy (Table 4.38). Only a few direct and indirect impacts will be discussed in this section. Direct economic impacts include impacts on per capita consumption, public sector revenue and expenditures, and regional economic stability (Jain et al., 2012). Indirect impacts include social externalities such as lifestyle effects, psychological effects, physiological effects, and community needs (Jain et al., 2012).

Mining has direct and indirect impacts (or social externalities) on the economy. The linkages described below provide indicators on how the economy is affected; however, taking into account all linkages is time consuming and costly. As an alternative to a complete economic analysis, a more narrow scope is recommended for determining economic impacts. The following four attributes are suggested:

- Per capita consumption
- Public sector revenue and expenditures

Table 4.38 Economic attributes affected by mining and mineral processing activities (Jain et al., 2012)

Economic attributes	Effect on economy	Source
Backward linkages	Per capita consumption can increase or decrease depending on the strength of the supply chain	Direct and indirect suppliers of equipment and services to a mine
Forward linkages	Regional economic stability can increase or decrease depending on whether or not forward linkages are used	Processing minerals prior to exportation or using minerals in country rather than exporting
Fiscal linkages	Public sector revenues and expenditures typically increase with mining	Income tax on those who work in the mine, tax on mineral profits, and payments to government for local goods and services
Consumption linkages	Per capita consumption can increase	Wages and profits associated with mining results in an increase in local business activity

- Regional economic stability
- Secondary and indirect impacts

These four attributes will provide an initial understanding of the economic effects of mining and mineral processing.

Direct Impacts

The direct economic impacts of mining and mineral processing can be measured through three primary categories:

- Per capita consumption
- Public sector revenue and expenditures
- Regional economic stability

For all direct impacts, it is important to accurately measure the baseline values in order to determine the economic effects of a mining project. Direct impacts are often projected over the lifetime of the mining project, where comparisons can be made between the cases where the mining project was not undertaken, and where it was undertaken. Variables may be measured on a yearly basis to determine estimated annual change, or on a function-by-function basis to determine how individual factors would be affected (Jain et al., 2012).

Per Capita Consumption

Per capita consumption is the yearly use of goods and services by each person, derived by dividing the quantity of goods and services used by the total population. This variable serves as a direct measure of personal economic well-being (Jain et al., 2012). Per capita consumption is affected by (Jain et al., 2012):

- Local employment
- Industrial expansion
- Construction

Any changes in local employment, construction, or industrial expansion will also affect the demand for local goods and services, in turn increasing or decreasing the amount of disposable income available. For example, new mining projects and industrial expansion will likely increase local employment, increasing the amount of disposable income and increasing per capita consumption (and economic well-being).

Per capita consumption is measured by (Jain et al., 2012):

- Inputs into mining, also known as backward linkages
- Direct employment changes, changes in average wages, and changes in the size of the labor force

- Changes in disposable income
- Changes in personal consumption

Changes in disposable income, personal consumption, employment, and wages are directly related to per capita consumption, but can be difficult to determine as other factors such as tax rates and filing bases, business failures and disruptions, dividends, interest, and ratios of wages to industry output, must be considered. Backward linkages for mining may include the supply of raw materials, retail goods, and mining equipment, but a ratio of mining inputs to outputs needs to be determined to accurately estimate backward linkages. Detrimental impacts of mining to per capita consumption can be mitigated by establishing direct linkages between the mining project and local businesses, ensuring the flow of money through the local economy (Jain et al., 2012).

Public Sector Revenue and Expenditures

Public sector revenue and expenditures is a measureable economic impact that results from mining. It is an expression of the annual per capita revenues and expenditures of local and state governments and associated agencies in a given region (Jain et al., 2012). Changes in public sector revenue and expenditures reflect the overall well-being of the public sector. Public sector revenue and expenditures are affected by (Jain et al., 2012):

- Employment
- Industrial and manufacturing sectors
- Acquisition and loss of real estate by agency activity

All these of these factors can have significant impacts on the economic, social, and physical conditions of a given area. For example, a new mining project and the jobs it creates would directly change personal income, which in turn would affect tax revenues and the cost of services, such as education, health, utilities, public welfare, and transportation. Changes in land usage and associated values may also affect revenues.

Variables for the measurement of public sector revenues and expenditures include (Jain et al., 2012):

- Changes in tax revenue
- Change in payment to local governments for goods or services (e.g., water or public utilities)
- Changes in local public expenses due to the addition of a mine or mining activity
- Change in personal income

- Damages to public facilities or other temporary or permanent costs that might result from mining project or activity

Determining the effect of a mining project on these variables is a highly complex process. Tax revenue will vary depending on local sales and income taxes, and other taxes such as the gasoline tax. For example, the gasoline tax alone may require knowledge of how increased personal income affects the number of vehicles in a given area, and how much gasoline is consumed as a result. Changes in public expenses (e.g., infrastructure, health, education) are usually assumed to be proportional to the change in personal income, which can be determined from changes in per capita consumption. Negative impacts on public sector revenues can be mitigated by providing input on the mining project to reduce costs to the local community by reducing demands on the public sectors services (e.g., repair of infrastructure), or by increasing direct and indirect payments to the local or regional government (Jain et al., 2012).

Regional Economic Stability

Regional economic stability can be defined as the speed and ease an economy returns to equilibrium after receiving a shock, or the ability to withstand severe fluctuations (Jain et al., 2012). Another definition is the degree of homogeneity of the region's economic activities in contributing to the gross regional product. The more diverse and closely related an economy is to growth areas of the national economy, the more stable it is likely to be. Regional economic stability is affected by (Jain et al., 2012):

- Any activity that results in some input or output relationship with a local business or individual (i.e., linkages) (Bloch and Owusu, 2012)
  - Backward linkages
  - Forward linkages
  - Fiscal linkages
  - Consumption linkages
- Direct purchases
- Indirect purchases through payroll

The severity of any change caused by an activity, such as mining, will be directly related to the dependence of the local economy on that specific sector. Economies that are highly dependent on one industry become vulnerable to all factors that affect that industry. Extreme examples of this include military or mining towns.

Measurements of regional economic stability include (Jain et al., 2012):

- Indications of effects include the percentage of total regional economic activity affected by activity
- Direct purchase of labor or other materials from the local economy

The two primary goals of regional economies are stability and growth. However, stability and growth are often incompatible, as high growth rates typically require some degree of specialization, which decreases stability. Negative impacts to economic stability can be mitigated by changing the distribution of demand for various economic outputs to closely represent national outputs (i.e., increasing economic diversity) or by increasing the demand for the output of high-growth industries (Jain et al., 2012). However, increasing economic diversity does not necessarily mean avoiding regional specialization.

Mining Specific Economic Impacts

Mining effects many economic attributes as previously noted. However, because of the ways these attributes are measured and affected, they are all interconnected. As such, this book discusses the economic attributes as "linkages." Backwards and forwards linkages in the local area, employment and wages, and the changes in tax revenues and money going to the local government have an effect on the condition of the economy. How a mining company interacts with the local communities affects how the economy responds to mining. Due to this fact, there can be positive or negative factors that result in increased or decreased economic activity.

The recent body of literature starting in the mid 1990's has found that when mining companies do not properly develop their interactions with local and national economies, a "resource curse" may develop. A resource curse is when countries that have abundant resources perform economically worse than their counterparts that do not have the same resource abundance (Zhang et al., 2008). More recent literature has questioned the existence of this alleged resource curse due to analytical issues with previous studies and found that the resource curse is more of an exception than a rule (Alexeev and Conrad, 2011; Brunnschweiler, 2008; Loayza et al., 2013). Since no definitive conclusions have been drawn regarding the existence of the resource curse, it will still be discussed in this section.

The resource curse is most strongly associated with point source resources, such as minerals and oil, and has been found to have a negative economic impact on countries or regions with "transition economies" (Alexeev and Conrad, 2011), defined as economies undergoing radical changes in the past 20 years, indicating that the existence of a resource curse might depend on existing economic conditions. Although the resource curse has been studied profusely, there are still uncertainties regarding whether certain factors contribute to the resource curse or result because of the curse. Several factors that have been associated with a resource curse or result from a resource curse are as follows:

- Poor institutional quality and institutional failures such as conflict, mismanagement, or corruption (Aragón and Rud, 2013; Brunnschweiler, 2008; Mideksa, 2013; Zhang et al., 2008)
  - Weak institutions allow resource profits to be spent in government consumption rather than investment (Brunnschweiler, 2008)
  - Resource abundance can result in armed conflict (Mideksa, 2013)
  - Revenue from natural resource endowments tends to sustain bad policies, which would otherwise not be sustainable without mining tax revenues (Mideksa, 2013)
- Weak linkages with other sectors of the economy
- Loss of learning by doing and other knowledge externalities associated with the more dynamic tradable sector (Mideksa, 2013)
- Rent seeking behavior (Stijns, 2006; Zhang et al., 2008)
- Revenues from natural resources may misallocate talent since such revenues are prone to rent seeking (Mideksa, 2013)
- Dutch Disease: the mechanism where an increase in revenues from natural resources will make a given nation's currency stronger compared to that of other nations (manifested in the appreciation of the exchange rate), resulting in the nation's other exports becoming more expensive for other countries to buy, making the manufacturing or agricultural sectors less competitive (Ebrahim-zadeh, 2003; Mideksa, 2013; Zhang et al., 2008)

If these factors are addressed by the mining companies, measures can be taken to reduce the likelihood of areas with mining developing a resource curse.

Economic Linkages

There is a negative correlation between natural resource abundance and consumption growth (Zhang et al., 2008). Many mining companies are perceived to not be connected to the local economy, although recent research has shown that mining is more deeply linked to local economies than expected (Bloch and Owusu, 2012). Aragon & Rud (2013) found that if there is a presence of strong backward linkages in a mining economy (i.e., if the mining industry uses local goods, services, and products from the surrounding area), then natural resources have the potential to be a blessing instead of a resource curse. The following discussion involves the different types of linkages relating mining to the economy.

Types of linkages between the mining industry and the economy include (Bloch and Owusu, 2012):

- *Forward linkages*: when one industry is producing raw input for another industry. In this case, the mines are producing ore or minerals for use in another industry. There are several ways that forward linkages can be used to increase economic advantage.
  - Minerals and metals can be processed prior to exportation
  - Minerals and metals can be used in the local economy rather than being exported (e.g., gold mined for local use as jewelry or sand mined for use in local construction)
- *Fiscal linkages*: how the government is financially connected to the mining industry.
  - State and federal taxation of income streams associated with the mining project
  - Policies that promote local procurement and employment may be more beneficial to local residents than increased public spending (Aragón and Rud, 2013)
  - When weak institutions allow resource profits to be spent in government consumption rather than investment, especially in countries with low levels of savings, the economy can be negatively impacted (Brunnschweiler, 2008)
- *Consumption linkages*: income in form of profits and wages that arise from commodity production (Bloch and Owusu, 2012).
  - Businesses in mining communities can be stimulated by and may depend upon profits and wages from mining
  - Those benefitting from this linkage include a range of formal and informal business activities – agriculture, animal husbandry, food, and housing
- *Backward linkages*: when activities are established or supported in order to supply inputs into the production of a commodity. The following tiers represent the many backward linkages that are developed to support the mining industry. Many of these linkages are from international sources, though mining companies are beginning to increase their use of local backward linkages (Aragón and Rud, 2013; Bloch and Owusu, 2012).
  - Tier 1: Direct suppliers
    - Engineering and service providers (e.g., project engineering companies)
    - Original equipment manufacturers (OEMs) (e.g., capital equipment)

- Consumables input suppliers (e.g., explosives, chemicals)
- Agents and distributors (e.g., pumps, bearings, vehicle parts)
- Tier 2: Indirect suppliers
  - Specialized engineering and services (e.g., electrical engineering, ventilation)
  - Component manufacturers
  - Manufacturers of standard components (e.g., cabling, electrical motor parts)
  - Manufacturers of specialized niche components (e.g., hoisting hooks, pinch valves)
  - Foundries and machine shops
  - Input providers (e.g., chemicals, steel products)
- Tier 3: Direct mining services
  - Surveying, geological and land use planning
  - Laboratory services
  - Drilling services
- Tier 4: Indirect producer services (which range from basic to sophisticated):
  - Finance, insurance, and real estate
  - Legal services
  - Transportation and logistics
  - Civil engineering
  - Environmental services
  - Construction and landscaping
  - Catering
  - Cleaning
  - Security

Indirect Impacts

Social externalities are a result of both positive and negative economic impacts from mining. The condition of local economies affects the lifestyles and health of local citizens. The four following categories are examples of externalities and indirect impacts caused by mining operations (Jain et al., 2012).

- Lifestyles
- Community needs
- Psychological needs
- Physiological needs

Because psychological and physiological needs are difficult to measure, this section will primarily focus on indirect impacts to lifestyles and community needs.

Lifestyles

Lifestyles refer to how people live and interact with each other, formally and informally. This can include social activities and standards of living of a population. Lifestyles are primarily affected by (Jain et al., 2012):

- Employment and wages
- Education levels
- Community development

Increases in employment, wages, and education will usually increase disposable income, resulting in a higher standard of living, usually resulting in increases in social and recreational activities. Community development is difficult to measure and will not be discussed in detail, as it depends on individual, formal and informal interactions, along with population demographics.

Employment and Wages

Several studies have argued that mining neither increases nor decreases local employment (Deller and Schreiber, 2013; Woods and Gordon, 2011). Deller & Schreiber (2013) noted that mining in an area resulted in an increase in the skills needed in workers (as opposed to historic mining activities that required low skilled workers) and an increase in wages; however, the study found that mining seemed detrimental to population growth. Woods & Gordon (2011) noted that this lack of correlation between mining (specifically coal mining in their study) and job creation meant that using mining activities to boost local employment is not reliable.

Jain et al. (2012) explains that large development projects such as mining operations, in practice, are loosely related to decreases in unemployment. New jobs requiring skilled workers are likely to be filled by currently employed workers looking to improve their job status, or the jobs will provide a stimulus for people not in the work force (e.g., a stay at home parent) to start working. In both of these cases, job openings will be filled (an increase in employment) without any decrease in unemployment in a local area. Skilled workers who travel to mining operations for employment will often base their families elsewhere, causing much of the income to leave the local economy, leaving the local community to bear the costs of the mine with few of the benefits (Reeson et al., 2012). Employment in other local sectors may also suffer due to staff shortages. Depending on the lifespan of the mine, businesses and services built to support growing mine communities may flourish in the short term but eventually suffer in the long term if mining revenue and jobs leave and the community cannot successfully transition back into a non-mining centered economy (Reeson et al., 2012).

Income inequality is a concern in local and regional areas that have mining activities (Fan et al., 2012; Loayza et al., 2013; Reeson et al., 2012; Zhang et al., 2008). Reeson et al. (2012) found that for men living in mining regions of Australia, income inequality initially increases with mining activity until a certain level of activity is reached (e.g., medium to high levels of mining employment in a community), at which more people are earning higher wages and income inequality decreases. The initial increase in inequality is primarily due to the difference in income between the mining sector and other low paying sectors in the community (O'Leary and Boettner, 2011). Income inequality among women increases as mining increases, due to the highly physical nature of many mining jobs. The study suggests that the highest levels of income inequality happen in areas with intermediate amounts of mining activity and inequality affects men and women differently. Another study by Grigoryan (2013) also found that while mining does enhance economic growth due to wage and profit distributions, it also increases inequality and poverty in local communities, possibly due to poor internalization of negative consequences (e.g., natural and local resource deterioration) that may unproportionally affect those of lower socioeconomic status. Yet another study, done by Loayza et al. (2013), finds that mining increases the standard of living and wages, and decreases poverty rates for producing districts. However, when looking at district level data, inequality increases and economic impacts decrease with greater geographic distance from the mine.

Education

Similar to the debate surrounding the existence of the resource curse, studies disagree on whether education is negatively impacted by mining. Some studies have found that increased resource abundance is linked with decreased spending in education (Brunnschweiler, 2008). This may be due to the fact that many mining jobs do not require as highly skilled or educated labor forces as other sectors, and thus, the workforce sees little benefit in more education (O'Leary and Boettner, 2011). This effect can be especially pronounced in communities dominated by the mining industry.

Contrasting studies have noted that high abundance of resources is associated with improved education (Stijns, 2006) and as a result, lower illiteracy rates (Loayza et al., 2013). Similarly, another study hypothesized that the resource curse can in fact be offset by higher education levels, making natural resource abundance a boon for countries with high human capital levels (Bravo-Ortega and De Gregorio, 2005). All studies used different measures of resource abundance, human capital, or level of education. To date, the effects of mining on education are inconclusive.

Community Needs

Community needs relate to the services that a community requires. These services are usually supplied by a local government and paid for by the community. Community needs can include housing, water supply, sewage disposal facilities, utilities, communications, recreational facilities, police, and fire protection (Jain et al., 2012). Community needs are affected by (Jain et al., 2012):

- Changes in population
- Changes in lifestyles

Mining and mineral extraction has long been connected to rapid population growth and decline, along with high demands on the public services (Weber et al., 2014). Increases in a population or drastic changes in lifestyles, especially due to temporary workforces and increases in wages brought about from mining operations, account for most of the changes in community needs.

Community needs are measured by (Jain et al., 2012):

- Changes in the population and characteristics of the population
- Changes in available housing
- Number of homes with connected utilities
- Size of the police and firefighting department
- Area of land dedicated for recreation

Most of these changes will occur rapidly due to an influx of people working at a mining operation, and again when workers leave after a mining operation closes. The issue of adequate housing remains a key issue in areas experiencing a mining boom. Without adequate housing expansion, massive influxes of workers can cause rent and property values to rapidly increase, preventing locals from obtaining affordable housing (Rolfe et al., 2007). In some cases, the lack of housing can force residents into substandard living conditions, homelessness, and temporary labor housing (Weber et al., 2014). These living conditions can place additional burdens on public services such as child protection services, foster care, and homeless shelters (Weber et al., 2014). Reports of benefits provided by increased employment, such as decreases in unemployment and food stamps, are difficult to confirm, as keeping track of demographic data is difficult during a boom (Weber et al., 2014).

Public services will have to expand to meet the needs of the new workforce, which can cause the community to incur debt in the process. However, once the temporary workforce leaves, the community will be left with the debt and services that are no longer needed, causing some to resist expansion. In some cases, the mining industry may donate to support local issues, especially those that directly affect the industry, such as road maintenance (Weber et al., 2014). However, costs associated with the expansion of law enforcement, emergency response, public health, and public works are often shouldered by the community. These costs are pronounced during the bust portion of a boom-bust cycle, when mining operations slow down or shut down all together.

Negative impacts to community needs can be mitigated by accounting for increases in population during the planning phases of a mining operation, and implementing alternative uses for services that are no longer needed when the population decreases (Jain et al., 2012). Communities should work with mining companies to determine the need for public services that will be brought about by a new mining operation. Building additional housing beforehand (whether temporary labor housing or traditional housing), obtain funding from mining companies, and scaling up public services to meet the expected demand can greatly minimize impacts to community needs.

Psychological Needs

Psychological needs encompass the emotional stability and security of the population and are affected by certain aspects of a mining project or activity (Jain et al., 2012). Aspects, such as motivation, behavior, and learning processes could be considered with psychological needs, but due to their difficultly in being related to changes in outside factors, are not included. Factors affecting psychological needs are likely to be the same factors that affect lifestyles. A new mining project may provide jobs increasing emotional stability and security, but may also push residents off of their land, decreasing stability. Impacts can be minimized by funding action plans for job placement after a mining operation has concluded in an area (Jain et al., 2012).

Physiological Needs

Physiological needs relate to the health and physical well-being of the population (Jain et al., 2012). Mining can negatively impact the health of nearby community members and mine workers. Mining activities such as land clearing, blasting, construction, and transportation can all threaten or harm human health. Other occupational hazards, such as exposure to coal dust or other mineral dusts can result in chronic respiratory diseases. The results of physiological needs can cause increases in the cost of health care, although implementing risk management systems and requiring personal protection equipment may minimize health impacts. Physiological needs of a population can also affect psychological needs as well as the entire economy if mining production is significantly affected (Jain et al., 2012).

REFERENCES

Abdul-Wahab, S., Marikar, F., 2012. The environmental impact of gold mines: pollution by heavy metals. Central Eur. J. Eng. 2, 304–313.

Agency for Toxic Substances and Disease Registry, 2005. Public Health Statement: Nickel CAS#: 7440-02-0. Department of Health and Human Services. Retrieved from: http://www.atsdr.cdc.gov/ToxProfiles/tp15-c1-b.pdf.

Agency for Toxic Substances and Disease Registry, 2010. Case Studies in Environmental Medicine: Lead Toxicity. Retrieved from: http://www.atsdr.cdc.gov/csem/lead/docs/lead.pdf.

Alexeev, M., Conrad, R., 2011. The natural resource curse and economic transition. Econ. Syst. 35, 445–461.

Anderson, J.O., Thundiyil, J.G., Stolbach, A., 2012. Clearing the air: a review of the effects of particulate matter air pollution on human health. J. Med. Toxicol. 8, 166–175.

Aneja, V.P., Isherwood, A., Morgan, P., 2012. Characterization of particulate matter (PM 10) related to surface coal mining operations in Appalachia. Atmos. Environ. 54, 496–501.

Aragón, F.M., Rud, J.P., 2013. Natural resources and local communities: evidence from a Peruvian gold mine. Am. Econ. J.: Econ. Pol. 5, 1–25.

Arey, J., Atkinson, R., 2006. Studies of the Atmospheric Chemistry of Volatile Organic Compounds and of Their Atmospheric Reaction Products. Retrieved from: http://arbis.arb.ca.gov/research/apr/past/99-330.pdf.

Argonne National Laboratory, 2008. Technology Demonstration for Reducing Mercury Emissions from Small-Scale Gold Refining Facilities. Retrieved from: http://www.ipd.anl.gov/anlpubs/2008/06/61757.pdf.

Attalla, M.I., Day, S.J., Lange, T., Lilley, W., Morgan, S., 2008. NO_x emissions from blasting operations in open-cut coal mining. Atmos. Environ. 42, 7874–7883.

Banks, J., 2012. Barriers and Opportunities for Reducing Methane Emissions from Coal Mines. Clean Air Task Force. Retrieved from: http://www.catf.us/resources/whitepapers/files/201209-Barriers_and_Opportunities_in_Coal_Mine_Methane_Abatement.pdf.

Barber, J.R., Crooks, K.R., Fristrup, K.M., 2010. The costs of chronic noise exposure for terrestrial organisms. Trends Ecol. Evol. 25, 180–189.

Bayne, E.M., Habib, L., Boutin, S., 2008. Impacts of chronic anthropogenic noise from energy – sector activity on abundance of songbirds in the boreal forest. Conserv. Biol. 22, 1186–1193.

Bell, F., Stacey, T., Genske, D., 2000. Mining subsidence and its effect on the environment: some differing examples. Environ. Geol. 40, 135–152.

Bempah, C.K., Ewusi, A., Obiri-Yeboah, S., Asabere, S.B., Mensah, F., Boateng, J., Voigt, H.-J., 2013. Distribution of arsenic and heavy metals from mine tailings dams at Obuasi Municipality of Ghana. Am. J. Eng. Res. 2, 61–70.

Bilotta, G.S., Brazier, R.E., 2008. Understanding the influence of suspended solids on water quality and aquatic biota. Water Res. 42, 2849–2861.

Bloch, R., Owusu, G., 2012. Linkages in Ghana's gold mining industry: challenging the enclave thesis. Resour. Polic. 37, 434–442.

Bowell, R., 2004. A review of sulfate removal options for mine waters. Proc. Mine Water, 75–88.

Boyd, R.S., 2010. Heavy metal pollutants and chemical ecology: exploring new frontiers. J. Chem. Ecol. 36, 46–58.

Bravo-Ortega, C., De Gregorio, J., 2005. The Relative Richness of the Poor? Natural Resources, Human Capital, and Economic Growth. World Bank Policy Research. Retrieved from: http://papers.ssrn.com/sol3/papers.cfm?abstract_id=648006.

Brunnschweiler, C.N., 2008. Cursing the blessings? Natural resource abundance, institutions, and economic growth. World Development 36, 399–419.

Canadian Council of Ministers of the Environment, 1999. Canadian Sediment Quality Guidelines for the Protection of Aquatic Life: Lead. Retrieved from: http://ceqg-rcqe. ccme.ca/download/en/239.

Canter, L., 1996. Environmental Impact Assessment, second ed. McGraw-Hill, Inc, New York.

Center for Research Information, 2004. Health Effects of Project SHAD Chemical Agent: Sulfur Dioxide. The National Academies. Retrieved from: https://www.iom.edu/ ~/media/Files/Report%20Files/2007/Long-Term-Health-Effects-of-Participation-in-Project-SHAD-Shipboard-Hazard-and-Defense/SULFURDIOXIDE.pdf.

Centers for Disease Control and Prevention, 2013. Lead Poisoning Investigation in Northern Nigeria. http://www.cdc.gov/onehealth/in-action/lead-poisoning.html.

Collett, R.S., Oduyemi, K., 1997. Air quality modelling: a technical review of mathematical approaches. Meteorol. Appl. 4, 235–246.

Deller, S., Schreiber, A., 2013. Mining and community economic growth. The Review of Regional Studies 42, 121–141.

Delmas, R., Serça, D., Jambert, C., 1997. Global inventory of NO_x sources. Nutrient Cycling in Agroecosystems 48, 51–60.

Department of Natural Resources (DNR), 2003. Mining Information Sheet – Protecting Groundwater at Metallic Mining Sites. Retrieved from: http://dnr.wi.gov/topic/ Mines/documents/gwa-pro.pdf.

Dooyema, C.A., Neri, A., Lo, Y.-C., Durant, J., Dargan, P.I., Swarthout, T., Biya, O., Gidado, S.O., Haladu, S., Sani-Gwarzo, N., 2011. Outbreak of fatal childhood lead poisoning related to artisanal gold mining in northwestern Nigeria, 2010. Environmental Health Perspect.

Ebrahim-zadeh, C., 2003. Back to Basics – Dutch Disease: Too Much Wealth Managed Unwisely, Finance and Development. International Monetary Fund.

Eckley, C.S., Gustin, M., Marsik, F., Miller, M.B., 2011. Measurement of surface mercury fluxes at active industrial gold mines in Nevada (USA). Sci. Tot. Environ. 409, 514–522.

Eisler, R., 1985a. Cadmium hazards to fish, wildlife, and invertebrates: a synoptic review. U.S. Fish and Wildlife Service. Retrieved from: https://www.pwrc.usgs.gov/eisler/CHR_2_Cadmium.pdf.

Eisler, R., 1985b. Selenium hazards to fish, wildlife, and invertebrates: a synoptic review. U.S. Fish and Wildlife Service. Retrieved from: https://www.pwrc.usgs.gov/eisler/CHR_5_Selenium.pdf.

Eisler, R., 1986. Chromium hazards to fish, wildlife, and invertebrates: a synoptic review. U.S. Fish and Wildlife Service. Retrieved from: https://www.pwrc.usgs.gov/eisler/CHR_6_Chromium.pdf.

Eisler, R., 1988. Arsenic hazards to fish, wildlife, and invertebrates: a synoptic review. U.S. Fish and Wildife Service. Retrieved from: https://www.pwrc.usgs.gov/eisler/CHR_12_Arsenic.pdf.

Eisler, R., 1993. Zinc hazards to fish, wildlife, and invertebrates: a synoptic review. U.S. Fish and Wildlife Service. Retrieved from: https://www.pwrc.usgs.gov/eisler/CHR_26_Zinc.pdf.

Eisler, R., 1998a. Copper hazards to fish, wildlife, and invertebrates: a synoptic review. U.S. Geological Survey. Retrieved from: https://www.pwrc.usgs.gov/eisler/CHR_33_Copper.pdf.

Eisler, R., 1998b. Nickel hazards to fish, wildlife, and invertebrates: a synoptic review. U.S. Geological Survey. Retrieved from: https://www.pwrc.usgs.gov/eisler/CHR_34_Nickel.PDF.

Environment Australia, 2002. Overview of Best Practice Environmental Management in Mining. Retrieved from: http://commdev.org/userfiles/files/1884_file_Overview_Best_Practice_Environmental_Management_in_Mining20051123111536.pdf.

Environment Protection Authority, 2004. Photochemical Smog – What It Means for Us. Retrieved from: http://www.epa.sa.gov.au/xstd_files/Air/Information%20sheet/info_photosmog.pdf.

Environmental Law Alliance Worldwide, 2010. Guidebook for Evaluating Mining Project EIAs, first ed. Retrieved from: http://www.elaw.org/files/mining-eia-guidebook/Full-Guidebook.pdf.

European Science Foundation, 2005. Volatile Organic Compounds in the Biosphere-Atmosphere System (VOCBAS). Retrieved from: http://www.esf.org/fileadmin/Public_documents/Publications/Volatile_Organic_Compounds_in_the_Biosphere-Atmosphere_System__VOCBAS_.pdf.

European Union, 2015. Air Quality Standards. http://ec.europa.eu/environment/air/quality/standards.htm.

Evuti, A.M., 2013. A synopsis on biogenic and anthropogenic volatile organic compounds emissions: hazards and control. Int. J. Eng. 2, 145.

Fan, R., Fang, Y., Park, S.Y., 2012. Resource abundance and economic growth in China. China Econ. Rev. 23, 704–719.

Flanagan, D.C., Gilley, J.E., Franti, T.G., 2007. Water Erosion Prediction Project (WEPP): development history, model capabilities, and future enhancement. Transactions of the ASABE 50, 1603–1612.

Francis, C.D., Kleist, N.J., Ortega, C.P., Cruz, A., 2012. Noise pollution alters ecological services: enhanced pollination and disrupted seed dispersal. Proc. R. Soc. B: Biol. Sci. 279, 2727–2735.

Gautam, S., Prusty, B.K., Patra, A.K., 2012. Pollution due to particulate matter from mining activities. Reciklaža i održivi razvoj 5, 53–58.

Ge, L., Chang, H.-C., Rizos, C., 2007. Mine subsidence monitoring using multi-source satellite SAR images. Photogrammetric Engineering & Remote Sensing 73, 259–266.

Gibb, H., O'Leary, K.G., 2014. Mercury exposure and health impacts among individuals in the artisanal and small-scale gold mining community: a comprehensive review. Environ. Health Perspect. 122, 667.

Global Methane Initiative, 2011. Coal Mine Methane: Reducing Emissions, Advancing Recovery and Use Opportunities. Retrieved from: https://www.globalmethane.org/documents/coal_fs_eng.pdf.

Global Water Intelligence, 2011. Water for Mining: Opportunities in Scarcity and Environmental Regulation. Retrieved from: http://www.globalwaterintel.com/client_media/uploaded/Water%20for%20Mining/GWI_Water_for_Mining_sample_pages-.pdf.

Golder Associates, 2011. Guidance Document on Water and Mass Balance Models for the Mining Industry. Retrieved from: http://www.env.gov.yk.ca/publications-maps/documents/mine_water_balance.pdf.

Govind, P., Madhuri, S., 2014. Heavy metals causing toxicity in animals and fishes. Res. J. Anim. Vet. Fishery Sci. 2, 17–23.

Gray, K.A., Finster, M.E., 1999. The Urban Heat Island, Photochemical Smog, and Chicago: Local Features of the Problem and Solution. Northwestern University, Department of Civil Engineering.

Greenpeace International, 2010. Mining Impacts.

Grigoryan, A., April 2013. In: The Impact of Mining Sector on Growth, Inequality, and Poverty: Evidence from Armenia, American University of Armenia Symposium, Emerging Issues in Environmental and Occupational Health,, pp. 22–23.

Gunson, A., Klein, B., Veiga, M., Dunbar, S., 2012. Reducing mine water requirements. J. Clean. Prod. 21, 71–82.

Haddad, N.M., Brudvig, L.A., Clobert, J., Davies, K.F., Gonzalez, A., Holt, R.D., Lovejoy, T.E., Sexton, J.O., Austin, M.P., Collins, C.D., 2015. Habitat fragmentation and its lasting impact on earth's ecosystems. Sci. Adv. 1, e1500052.

Hagen, L., 2010. Erosion by wind: modeling. Encyclopedia Soil Science, 1–4.

Harris, J., Hunter, D., Birch, P., Short, K., 1987. Vesicular arbuscular mycorrhizal populations in stored topsoil. Transactions of the British Mycological Society 89, 600–603.

Head, H.J., Kissell, F.N., 2006. Chapter 13—Methane control in metal/nonmetal mines. In: Handbook for Methane Control in Mining National Institute for Occupational Safety and Health.

Hilson, G., 2002. An overview of land use conflicts in mining communities. Land Use Policy 19, 65–73.

Hughes, L., 2005. NSW Scientific Committee – Final Determination: Alteration of Habitat Following Subsidence due to Longwall Mining – Key Threatening Process Listing. New South Wales Government. Retrieved from: http://www.environment.nsw.gov.au/determinations/LongwallMiningKtp.htm.

Human Rights Watch, 2011. A Heavy Price: Lead Poisoning and Gold Mining in Nigeria's Zamfara State. Retrieved from: http://www.hrw.org/sites/default/files/related_material/Nigeria_0212.pdf.

Intergovernmental Panel on Climate Change, 2006. 2006 IPCC Guidelines for National Greenhouse Gas Inventories United Nations. Retrieved from: http://www.ipcc-nggip.iges.or.jp/public/2006gl/pdf/0_Overview/V0_1_Overview.pdf.

International Atomic Energy Agency, 2005. Guidebook on Environmental Impact Assessment for In Situ Leach Mining Projects. Retrieved from: http://www-pub.iaea.org/MTCD/publications/PDF/te_1428_web.pdf.

International Council on Mining & Metals (ICMM), 2006. Good Practice Guidance for Mining and Biodiversity. Retrieved from: http://www.icmm.com/document/13.

International Council on Mining & Metals (ICMM), 2012. Water Management in Mining: A Selection of Case Studies. Retrieved from: https://www.icmm.com/document/3660.

Jain, R., Urban, L., Stacey, G.S., Balbach, H., Webb, M.D., 2012. Handbook of Environmental Engineering Assessment: Strategy, Planning, and Management. Elsevier, Waltham, MA.

Jennings, S.R., Neuman, D.R., Blicker, P.S., 2008. Acid Mine Drainage and Effects on Fish Health and Ecology: A Review. Reclamation Research Group. Retrieved from: http://www.pebblescience.org/pdfs/Final_Lit_Review_AMD.pdf.

Johnson, D.B., Hallberg, K.B., 2005. Acid mine drainage remediation options: a review. Sci. Tot. Environ. 338, 3–14.

Kight, C.R., Swaddle, J.P., 2011. How and why environmental noise impacts animals: an integrative, mechanistic review. Ecol. Lett. 14, 1052–1061.

Klimont, Z., Smith, S.J., Cofala, J., 2013. The last decade of global anthropogenic sulfur dioxide: 2000–2011 emissions. Environ. Res. Lett. 8, 014003.

Lashgari, A., Johnson, C., Kecojevic, V., Lusk, B., Hoffman, J.M., 2013. NO_x emission of equipment and blasting agents in surface coal mining. Min. Engin. 65, 34–41.

Leading Practice Sustainable Development Program for the Mining Industry, 2006. Mine Rehabilitation. Australian Department of Industry, Tourism and Resources. Retrieved from: http://www.industry.gov.au/resource/Documents/LPSDP/LPSDP-MineRehabilitationHandbook.pdf.

Leading Practice Sustainable Development Program for the Mining Industry, 2007a. Biodiversity Management. Australian Department of Industry, Tourism and Science. Retrieved from: http://www.industry.gov.au/resource/Documents/LPSDP/LPSDP-BiodiversityHandbook.pdf.

Leading Practice Sustainable Development Program for the Mining Industry, 2007b. Managing Acid and Metalliferous Drainage. Australian Department of Industry, Tourism and Resources. Retrieved from: http://www.industry.gov.au/resource/Documents/LPSDP/LPSDP-AcidHandbook.pdf.

Leading Practice Sustainable Development Program for the Mining Industry, 2008. Water Management. Australian Department of Resources, Energy and Tourism. Retrieved from: http://www.industry.gov.au/resource/Documents/LPSDP/LPSDP-WaterHandbook.pdf.

Leading Practice Sustainable Development Program for the Mining Industry, 2009. Airborne Contaminants, Noise and Vibration. Australian Department of Resources, Energy and Tourism. Retrieved from: http://www.industry.gov.au/resource/Documents/LPSDP/AirborneContaminantsNoiseVibrationHandbook_web.pdf.

Li, Z., Ma, Z., van der Kuijp, T.J., Yuan, Z., Huang, L., 2014. A review of soil heavy metal pollution from mines in China: pollution and health risk assessment. Sci. Tot. Environ. 468–469, 843–853.

Loayza, N., Mier y Teran, A., Rigolini, J., 2013. Poverty, Inequality, and the Local Natural Resource Curse. Discussion Paper Series, Forschungsinstitut zur Zukunft der Arbeit, Retrieved from: http://elibrary.worldbank.org/doi/pdf/10.1596/1813-9450-6366.

Lottermoser, B.G., 2010. Mine Wastes: Characterization, Treatment and Environmental Impacts, third ed. Springer, New York.

Madden, F., Camcigil, B., 2000. TRI Toolkit. Mineral Policy Center. Retrieved from: http://www.earthworksaction.org/files/publications/toolkit.pdf.

Mainiero, R.J., Harris, M.L., Rowland, J.H., 2007. Dangers of toxic fumes from blasting. In: Proceedings of the Annual Conference on Explosives and Blasting Technique. Int. Soc. Explos. Eng.

Martin, S., Griswold, W., 2009. Human Health Effects of Heavy Metals. Center for Hazardous Substance Research. Kansas State University. Retrieved from: https://www.engg.ksu.edu/CHSR/outreach/resources/docs/15HumanHealthEffectsofHeavyMetals.pdf.

Maupin, M.A., Kenny, J.F., Hutson, S.S., Lovelace, J.K., Barber, N.L., Linsey, K.S., 2014. Estimated Use of Water in the United States in 2010. Circular 1405. U.S. Geological Survey. Retrieved from: http://pubs.usgs.gov/circ/1405/pdf/circ1405.pdf.

Mideksa, T.K., 2013. The economic impact of natural resources. J. Environ. Econ. Manage. 65, 277–289.

Mielke, R.E., Pace, D.L., Porter, T., Southam, G., 2003. A critical stage in the formation of acid mine drainage: colonization of pyrite by *Acidithiobacillus ferrooxidans* under pH-neutral conditions. Geobiol. 1, 81–90.

Milieu Ltd., Danish National Environmental Research Institute, Center for Clean Air Policy, 2004. Assessment of the Effectiveness of European Air Quality Policies and Measures. Retrieved from: http://ec.europa.eu/environment/archives/cafe/activities/pdf/task_3_2_general.pdf.

Mine Safety and Health Administration (MSHA), 1997. Controlling Mercury Hazards in Gold Mining: A Best Practices Toolbox. U.S. Department of Labor. Retrieved from: http://www.msha.gov/S&HINFO/MERCURY/HGTB.pdf.

Mine Subsidence Engineering Consultants, 2007. Introduction to Longwall Mining and Subsidence. Retrieved from: http://www.minesubsidence.com/index_files/files/Intro_Longwall_Mining_and_Subs.pdf.

Mineral Policy Center, 2000. Cyanide Leach Mining Packet. Retrieved from: http://www.earthworksaction.org/files/publications/Cyanide_Leach_Packet.pdf.

Mummey, D.L., Stahl, P.D., Buyer, J.S., 2002. Soil microbiological properties 20 years after surface mine reclamation: spatial analysis of reclaimed and undisturbed sites. Soil Biol. Biochem. 34, 1717–1725.

National Academy of Engineering (NAE), 2010. Grand challenges for Earth Resources Engineering. National Academy of Engineering. Retrieved from: https://www.nae.edu/File.aspx?id=106323.

National Pollutant Inventory, 1999. Emission Estimation Technique Manual for Mining and Processing of Non-Metallic Minerals. Australian Government Department of the Environment. Retrieved from: http://www2.unitar.org/cwm/publications/cbl/prtr/pdf/cat5/fnonmeta.pdf.

National Pollutant Inventory, 2012. Emission Estimation Technique Manual for Fuel and Organic Liquid Storage. Version 3.3. Australian Government Department of the Environment. Retrieved from: http://www.npi.gov.au/system/files/resources/5d886b0c-d392-4c04-c91d-a3a099bc0988/files/fols.pdf.

National Toxicity Program, 2012. NTP Monograph: Health Effects of Low-Level Lead. U.S. Department of Health and Human Services. Retrieved from: http://ntp.niehs.nih.gov/ntp/ohat/lead/final/monographhealtheffectslowlevellead_newissn_508.pdf.

Nevada Mining Association, 2010. Air Quality Issues and Policy. http://www.nevadamining.org/issues_policy/air_quality.php.

O'Leary, S., Boettner, T., 2011. Booms and Busts: The Impact of West Virginia's Energy Economy. West Virginia Center on Budget & Policy.

Pennsylvania Department of Environmental Protection, 2012. Technical Guide to Mine Subsidence. Retrieved from: http://www.dep.state.pa.us/msihomeowners/technicalguidetoms.html.

Pirrone, N., Cinnirella, S., Feng, X., Finkelman, R., Friedli, H., Leaner, J., Mason, R., Mukherjee, A., Stracher, G., Streets, D., 2010. Global mercury emissions to the atmosphere from anthropogenic and natural sources. Atmos. Chem. Phys. 10, 5951–5964.

Plante, B., Bussière, B., Benzaazoua, M., 2012. Static tests response on 5 Canadian hard rock mine tailings with low net acid-generating potentials. J. Geochem. Explor. 114, 57–69.

Plumlee, G.S., Logsdon, M.J., Filipek, L.H., 1999. The Environmental Geochemistry of Mineral Deposits. Pacific Sect. Soc. Econ. Geol.

Pourrut, B., Shahid, M., Dumat, C., Winterton, P., Pinelli, E., 2011. Lead uptake, toxicity, and detoxification in plants. Springer Rev. Environ. Contam. Toxicol. 213, 113–136.

Queensland Government, 2011. Queensland Guidance Note QGN 20 v. 3: Management of Oxides of Nitrogen in Open Cut Blasting. Retrieved from: https://www. oricaminingservices.com/uploads/Bulk%20Systems/QGN-mgmt-oxides-nitrogen.pdf.

Reed, W.R., 2005. Information Circular 9478: Significant Dust Dispersion Models for Mining Operations. Centers for Disease Control and Prevention. Retrieved from: http://www.cdc.gov/niosh/mining/UserFiles/works/pdfs/2005-138.pdf.

Reeson, A.F., Measham, T.G., Hosking, K., 2012. Mining activity, income inequality and gender in regional Australia. Aust. J. Agric. Res. Econ. 56, 302–313.

Rolfe, J., Miles, B., Lockie, S., Ivanova, G., 2007. Lessons from the Social and Economic Impacts of the Mining Boom in the Bowen Basin 2004–2006.

Safe Drinking Water Foundation, 2005. Mining and Water Pollution. Retrieved from: http://www.safewater.org/PDFS/resourcesknowthefacts/Mining+and+Water+Pollution.pdf.

Sahu, L., 2012. Volatile organic compounds and their measurements in the troposphere. Curr. Sci. 102, 1645–1649.

Schueler, V., Kuemmerle, T., Schröder, H., 2011. Impacts of surface gold mining on land use systems in Western Ghana. Ambio 40, 528–539.

Skousen, J., Simmons, J., McDonald, L.M., Ziemkiewicz, P., 2002. Acid-base accounting to predict post-mining drainage quality on surface mines. J. Environ. Qual. 31, 2034–2044.

Solomon, F., 2008. Impacts of metals on aquatic ecosystems and human health. Environ. Communities 14–19.

Spiegel, S.J., Veiga, M.M., 2010. International guidelines on mercury management in small-scale gold mining. J. Clean. Prod. 18, 375–385.

Stahl, P.D., Williams, S.E., Christensen, M., 1988. Efficacy of native vesicular-arbuscular mycorrhizal fungi after severe soil disturbance. New Phytologist 110, 347–354.

Stijns, J.-P., 2006. Natural resource abundance and human capital accumulation. World Develop. 34, 1060–1083.

Telmer, K., Veiga, M.M., 2008. World Emissions of Mercury from Small Scale Artisanal Gold Mining and the Knowledge Gaps about Them. GMP Presentation, Rome, Italy.

Todd, D.K., Mays, L.W., 2005. Groundwater Hydrology, third ed. Wiley.

Tong, S., von Schirnding, Y.E., Prapamontol, T., 2000. Environmental lead exposure: a public health problem of global dimensions. Bull. World Health Organization 78, 1068–1077.

U.S. Department of Agriculture, 2000. Heavy Metal Soil Contamination. Retrieved from: http://www.nrcs.usda.gov/Internet/FSE_DOCUMENTS/nrcs142p2_053279.pdf.

U.S. Energy Information Administration, 2013. International Energy Statistics: Total Primary Coal Production. Retrieved from: http://www.eia.gov/cfapps/ipdbproject/IEDIndex3.cfm?tid=1&pid=7&aid=1.

U.S. Environmental Protection Agency (U.S. EPA), 1994. Technical Document – Acid Mine Drainage Prediction. Retrieved from: http://water.epa.gov/polwaste/nps/upload/amd.pdf.

U.S. Environmental Protection Agency (U.S. EPA), 1995a. Compliation of Air Pollutant Emission Factors: Volume I: Stationary Point and Area Sources – Leadbearing Ore Crushing And Grinding. Retrieved from: http://www.epa.gov/ttn/chief/ap42/ch12/final/c12s18.pdf.

U.S. Environmental Protection Agency (U.S. EPA), 1995b. Compliation of Air Pollutant Emission Factors: Volume I: Stationary Point and Area Sources – Primary Copper Smelting. Retrieved from: http://www.epa.gov/ttn/chief/ap42/ch12/final/c12s03.pdf.

U.S. Environmental Protection Agency (U.S. EPA), 1995c. Compliation of Air Pollutant Emission Factors: Volume I: Stationary Point and Area Sources – Primary Lead Smelting. Retrieved from: http://www.epa.gov/ttn/chief/ap42/ch12/final/c12s06.pdf.

U.S. Environmental Protection Agency (U.S. EPA), 1995d. Compliation of Air Pollutant Emission Factors: Volume I: Stationary Point and Area Sources – Western Surface Coal Mining. Retrieved from: http://www.epa.gov/ttnchie1/ap42/ch11/final/c11s09.pdf.

U.S. Environmental Protection Agency (U.S. EPA), 1995e. Compliation of Air Pollutant Emission Factors: Volume I: Stationary Point and Area Sources – Zinc Smelting. Retrieved from: http://www.epa.gov/ttn/chief/ap42/ch12/final/c12s07.pdf.

U.S. Environmental Protection Agency (U.S. EPA), 1997a. EPA's National Hardrock Mining Framework: Appendix B Potential Environmental Impacts of Hardrock Mining. Retrieved from: http://www.epa.gov/aml/policy/app_b.pdf.

U.S. Environmental Protection Agency (U.S. EPA), 1997b. Mercury Study Report to Congress Volume VI: An Ecological Assessment for Anthropogenic Mercury Emissions in the United States. Retrieved from: http://www.epa.gov/ttn/oarpg/t3/reports/volume6.pdf.

U.S. Environmental Protection Agency (U.S. EPA), 2002. Supplemental Guidance for Developing Soil Screening Levels for Superfund Sites. Retrieved from: http://www.epa.gov/superfund/health/conmedia/soil/index.htm.

U.S. Environmental Protection Agency (U.S. EPA), 2009a. Drinking Water Contaminants. http://water.epa.gov/drink/contaminants/index.cfm.

U.S. Environmental Protection Agency (U.S. EPA), 2009b. Ventilation Air Methane (VAM) Utilization Technologies. Retrieved from: http://www.epa.gov/cmop/docs/vam_technology.pdf.

U.S. Environmental Protection Agency (U.S. EPA), 2011. The Benefits and Costs of the Clean Air Act from 1990 to 2020: Summary Report. Retrieved from: http://www.epa.gov/air/sect812/feb11/summaryreport.pdf.

U.S. Environmental Protection Agency (U.S. EPA), 2012a. Effects of Acid Rain. http://www.epa.gov/acidrain/effects/index.html.

U.S. Environmental Protection Agency (U.S. EPA), 2012b. Global Anthropogenic Non-CO_2 Greenhouse Gas Emissions: 1990–2030. Retrieved from: http://www.epa.gov/climatechange/Downloads/EPAactivities/EPA_Global_NonCO2_Projections_Dec2012.pdf.

U.S. Environmental Protection Agency (U.S. EPA), 2013. Case Study: Methane Recovery at Non-coal Mines. Retrieved from: http://www.epa.gov/cmop/docs/CMOP-Noncoal%20Flyer.pdf.

U.S. Environmental Protection Agency (U.S. EPA), 2014a. Clean Water Act (CWA). http://www.epa.gov/agriculture/lcwa.html.

U.S. Environmental Protection Agency (U.S. EPA), 2014b. Coalbed Methane Outreach Program: Promoting CMM Recovery and Use. Retrieved from: http://www.epa.gov/coalbed/docs/CMOP-General-Flyer.pdf.

U.S. Environmental Protection Agency (U.S. EPA), 2014c. National Ambient Air Quality Standards (NAAQS). http://www.epa.gov/air/criteria.html.

U.S. Environmental Protection Agency (U.S. EPA), 2014d. National Recommended Water Quality Criteria. http://water.epa.gov/scitech/swguidance/standards/criteria/current/index.cfm.

U.S. Environmental Protection Agency (U.S. EPA), 2014e. Particulate Matter (PM): Health. http://www.epa.gov/pm/health.html.

U.S. Environmental Protection Agency (U.S. EPA), 2014f. Technology Transfer Network Support Center for Regulatory Atmospheric Modeling: Preferred/Recommended Models. http://www.epa.gov/ttn/scram/dispersion_prefrec.htm#rec.

U.S. Environmental Protection Agency (U.S. EPA), 2015a. 2011 National Emissions Inventory. http://www.epa.gov/ttn/chief/net/2011inventory.html.

U.S. Environmental Protection Agency (U.S. EPA), 2015b. National Greenhouse Gas Emissions Data: Draft Inventory of U.S. Greenhouse Gas Emissions and Sinks: 1990–2013. Retrieved from: http://www.epa.gov/climatechange/pdfs/usinventoryreport/US-GHG-Inventory-2015-Chapter-3-Energy.pdf.

U.S. Environmental Protection Agency (U.S. EPA), 2015c. National Summary of Lead Emissions. In: http://www.epa.gov/cgi-bin/broker?_service=data&_program=dataprog.national_2. sas&_debug=0&sector=Industrial%20Processes&pol=7439921&polchoice=Pb.

U.S. Geological Survey (USGS), 2000. Coal-Bed Methane: Potential and Concerns. Retrieved from: http://pubs.usgs.gov/fs/fs123-00/fs123-00.pdf.

U.S. National Library of Medicine, 2014. Volatile Organic Compounds (VOCs). http://toxtown.nlm.nih.gov/text_version/chemicals.php?id=31.

United Nations Environment Programme, 2013. Global Mercury Assessment 2013: Sources, Emissions, Releases and Environmental Transport. United Nations. Retrieved from: http://www.unep.org/PDF/PressReleases/GlobalMercuryAssessment2013.pdf.

Weber, B.A., Geigle, J., Barkdull, C., 2014. Rural North Dakota's oil boom and its impact on social services. Social work 59, 62–72.

Woods, B.R., Gordon, J.S., 2011. Mountaintop removal and job creation: exploring the relationship using spatial regression. Ann. Assoc. Ame. Geogr. 101, 806–815.

World Bank Group, 1998. Pollution Prevention and Abatement Handbook: Sulfur Oxides. Retrieved from: http://www.ifc.org/wps/wcm/connect/5cb16d8048855 c248b24db6a6515bb18/HandbookSulfurOxides.pdf?MOD=AJPERES.

World Health Organization, 2007. Exposure to Mercury: A Major Public Health Concern. Retrieved from: http://www.who.int/phe/news/Mercury-flyer.pdf.

World Health Organization, 2010. Exposure to Air Pollution: A Major Public Health Concern. Retrieved from: http://www.who.int/ipcs/features/air_pollution.pdf.

World Health Organization, 2011. Guidelines for Drinking-Water Quality, fourth ed. Retrieved from: http://whqlibdoc.who.int/publications/2011/9789241548151_eng. pdf?ua=1.

World Health Organization, 2013. Health Effects of Particulate Matter: Policy Implications for Countries in Eastern Europe, Caucasus and Central Asia. Retrieved from: http://www.euro.who.int/__data/assets/pdf_file/0006/189051/Health-effects-of-particulate-matter-final-Eng.pdf.

Wu, Q., Wang, S., Zhang, L., Song, J., Yang, H., Meng, Y., 2012. Update of mercury emissions from China's primary zinc, lead and copper smelters, 2000–2010. Atmos. Chem. Phy. 12, 11153–11163.

Wuana, R.A., Okieimen, F.E., 2011. Heavy metals in contaminated soils: a review of sources, chemistry, risks and best available strategies for remediation. ISRN Ecol. 2011.

Yang, Y., 2007. Coal Mining and Environmental Health in China. China Environmental Forum. Retrieved from: http://www.circleofblue.org/waternews/wp-content/uploads/2011/03/coalmining_april2.pdf.

Yirenkyi, S., 2008. Surface Mining and Its Socio-Economic Impacts and Challenges. The Southern African Institute of Mining and Metallurgy. Retrieved from: http://www.saimm.co.za/Conferences/SurfaceMining2008/181-202_Yirenkyi.pdf.

Younger, P.L., Wolkersdorfer, C., 2004. Mining impacts on the fresh water environment: technical and managerial guidelines for catchment scale management. Mine Water Environ. 23, s2–s80.

Zhang, X., Xing, L., Fan, S., Luo, X., 2008. Resource abundance and regional development in China. Econ. Trans. 16, 7–29.

CHAPTER 5

Environmental Monitoring

5.1 INTRODUCTION

The monitoring process is essential to the implementation of and progress toward clean and sustainable production in the mining industry. Monitoring allows the collection of data before, during, and after mining activities to analyze and quantify environmental impacts. Similar to the cyclical process for evaluating environmental performance, the monitoring process is also cyclical. Incorporating monitoring into environmental management systems enables the process of monitoring, analyzing, reviewing, and improving, whereby promoting sustainable development (Figure 5.1).

Once the monitoring program has been implemented, data collected needs to be analyzed and the results used to improve performance and the monitoring process. These specifics will be discussed in this chapter.

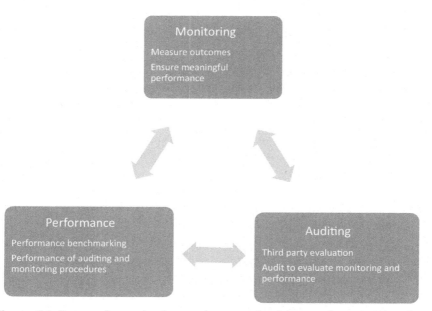

Figure 5.1 Process for evaluating environmental mining performance (Leading Practice Sustainable Development Program for the Mining Industry, 2009b).

Environmental Impact of Mining and Mineral Processing
ISBN 978-0-12-804040-9
http://dx.doi.org/10.1016/B978-0-12-804040-9.00005-X

The process of creating a monitoring network and associated procedures is as follows (Leading Practice Sustainable Development Program for the Mining Industry, 2009b):

1. Monitoring plan and design
2. Implementation
3. Rigorous review of collected data
4. Data interpretation
5. Adjustments and improvements to monitoring techniques and mining processes

Monitoring is used throughout the mining industry to (Environment Canada, 2011; Leading Practice Sustainable Development Program for the Mining Industry, 2009b):

- Establish baseline conditions
- Determine contaminants and toxic substances emitted, discharged, or resulting from mining operations
- Determine cumulative effects of land use alterations, discharges, contaminants, and substances on the environment
- Assess processes and quality assurance plans
- Provide information to guide future adjustments to mining processes to decrease environmental impacts
- Validate models
- Support decision making by governments and stakeholders
- Promote transparency through accessible and comparable data
- Comply with local and federal regulations

There are currently deficiencies in monitoring practices that decrease the effectiveness of monitoring programs. Without accurate and precise monitoring data or interpretation, the likelihood of finding and correcting environmental impacts decreases. Some common deficiencies include (Leading Practice Sustainable Development Program for the Mining Industry, 2009b):

- Lack of clearly defined objectives and purpose for monitoring programs that can lead to unsatisfactory outcomes, wasted resources, and potential conflict with stakeholders
- Data not analyzed or analyses not used to enable continuous improvement
- Performance measures being too narrow or fail to include adequate socioeconomic and environmental perspectives

- Inappropriate levels of public reporting, meaning that the purpose and context of monitoring and auditing are not clearly understood
- Monitoring plans being designed to only meet regulatory requirements and not address actual problems
  - Timeframes for review structured to meet regulatory requirements, but do not address problems
  - Annual monitoring reports used for regulatory compliance requirements only, and do not focus on performance or proactive problem solving
- Inadequate use of risk assessment methods to provide proper quality assurance/quality control (QA/QC)
- Monitoring plans not being flexible or adaptable to changes in mine life-cycle because the original plans focused mainly on start-up issues
- Adequate baseline data for managing long-term issues not obtained at the beginning of the project.

Many of these deficiencies can be addressed through careful planning and implementation.

5.2 DESIGN OF MONITORING PLANS

Monitoring plans should be designed during the feasibility phase of mine planning and should include procedures for the entire life of the mine. Baseline monitoring and impact monitoring should be included in mine monitoring plans for all time frames (e.g., an annual plan, a medium-term (3–5 years) plan, and a life-of-mine plan).

One common theme that should be incorporated into all monitoring plans is the need to be proactive. An effective monitoring program needs to go beyond regulatory compliance monitoring and focus on key parameters for the early detection of mine trends before issues become problems. Similarly, a monitoring program should not be limited to measuring the parameters that are in environmental or regulatory permits, but should incorporate other parameters that have been noted as areas of risk over the lifetime of the mine. Risk assessments play an essential role in monitoring design. The assessments must be based on sound science and validated knowledge of environmental sensitivities and environmental and stakeholder responses. In addition, the monitoring plan needs to be able to be adapted to changes in mining activity to adequately manage liability. The following parameters are a summary of considerations for monitoring planning (Leading Practice Sustainable Development Program for the Mining Industry, 2009b):

- Develop specific objectives to include in the monitoring plan to provide direction. Objectives may be based on the following considerations:
  - Primary purpose for monitoring (e.g., permits, reduce environmental impacts, public image, etc.)
  - End users of the monitoring data
  - Data uses and possible analyses
  - Parameters and conditions to be monitored and for what time period
  - Resolution and accuracy of the data
  - Monitoring and sampling methods
  - Monitoring locations and sampling sites
  - Monitoring time period and frequency
  - Data management, maintenance, and presentation
- Use risk assessments to identify areas of environmental and ecologic risk over the life of the mine and incorporate response to these risks in the monitoring plan
- Design the monitoring plan to have short-term and long-term relevancy with respect to the life of the mine
- Incorporate adaptability and flexibility to meet changing mining activities (Leading Practice Sustainable Development Program for the Mining Industry, 2009b)
  - Adapt monitoring and data collection to changes in mine operations and plans
  - Changes in the type of mining or in the ore that is mined and processed on site
  - Events that may cause the company to adjust assumptions on which planning has been based and on which risk was assessed
  - A significant incident at a similar mine site or in the same region
  - Changes within nearby communities due to mine changes throughout its life-cycle

Risks Associated with Environmental Monitoring Systems

When designing a monitoring program, there are two types of risk that can affect the design. The first risk involves the chance that environmental issues might arise during the mining process and seriously degrade environmental conditions. The second type of risk is specific to monitoring programs and includes any errors or mistakes that might be inherent to monitoring (Leading Practice Sustainable Development Program for the Mining Industry, 2009b).

To manage and asses risk related to environmental impacts, the following parameters must be considered in the monitoring plan (Leading Practice Sustainable Development Program for the Mining Industry, 2009b):

- Identify legal and regulatory requirements
- Use the company risk management framework (also known as a "risk register") to identify potentially significant environmental risks in order to develop and apply control measures
- Develop preventative and mitigation procedures for risks identified in the risk assessment

After defining these parameters, a risk assessment system can be developed that is sensitive enough to detect risks and apply mitigation measures before issues escalate. Appropriate sensitivity parameters and endpoints should be selected according to identified risks.

The second type of risk regarding the actual monitoring process and data collection may include the following (Leading Practice Sustainable Development Program for the Mining Industry, 2009b):

- Baseline monitoring is not carried out over a sufficiently representative time period or location to provide good quality data upon which to compare possible environmental impacts
- Monitoring installations are destroyed by vandalism, fire, flood, or feral animals, causing loss of data at critical times
- Databases used to manage and interpret data change over time and old data becomes irretrievable
- Personnel who understand the critical elements of a monitoring program do not document procedures and, when they leave the company, new personnel are unable to manage the monitoring system to the standard required
- Monitoring data are reviewed annually but not over the life of a project, causing cumulative impacts to go undetected
- Monitoring focuses on indirect measures of impact and therefore fails to detect the impacts that need to be measured

The frequency or intensity of these risks occurring should be incorporated into the monitoring design process. Control of baseline monitoring activities and installations should be covered in QA/QC protocols. Issues pertaining to longevity, continuity, and data analyses should be covered by the various short-term, medium-term, and long-term monitoring plans.

Baseline Monitoring Design

Obtaining accurate and precise baseline data is extremely important to the success of monitoring programs. Baseline data is used for comparison throughout the entire life of the mine to determine environmental impacts and must provide enough data to enable assessment of risk for each parameter in each stage of mine operation. Baseline monitoring should start in the pre-feasibility stage of mine planning and include all relevant environmental, social, and economic issues identified in risk planning (Leading Practice Sustainable Development Program for the Mining Industry, 2009b). Risk planning, as discussed above, is very important as it dictates the amount and extent of the baseline monitoring. Baseline monitoring should continue throughout the life of the mine. Comparing baseline data to impact data can provide insight into the variability caused by non-mining impacts and pre-existing conditions.

Baseline studies must be able to gather the necessary data to support predictions of potential impacts and determine strategies for mitigating impacts. The major objectives of an environmental baseline study should be to gather enough data in order to (British Columbia Ministry of Environment, 2012):

- Characterize risks to water, air, and land quality
- Characterize socioeconomic risks
- Determine impact mechanisms and pathways
- Establish relevant and safe thresholds for parameters indicative of ecosystem health
- Facilitate design of water and environmental quality monitoring program

One of the first steps in developing a baseline monitoring plan is to review existing information. Site information can be obtained from sources such as peer reviewed literature, reports from government agencies and research institutes, government monitoring datasets, geological surveys, long-term weather and climate monitoring stations, water quality monitoring stations, well records, photographs, aerial and satellite imagery, local licenses, maps, and various industries operating in the area of interest. Careful review of existing information can help determine the need for further monitoring and locations where monitoring may be most important.

Site selection for baseline monitoring varies depending on the type of monitoring. Sites should be reasonably accessible to allow for periodic maintenance and should be widespread to cover the entire area impacted by a mining project. Monitoring locations should be based on needs specified in the impact monitoring design plan. Sites up-gradient and down-gradient from the mining project should be selected for air quality, hydrology and water quality, and local flora and fauna to determine baseline conditions for any changes that may result from the project.

When determining baseline variables to be measured, it is important to understand the type of mining project in consideration. Some operations may pose a greater risk of contamination for specific variables (e.g., heavy metals), than others. For example, a uranium mine should place a high priority on radionuclide monitoring throughout baseline monitoring and well past remediation. However, other mines may not necessarily need to be concerned with radionuclides, and once baseline levels are established, if no increases are seen during operation, radionuclides can become candidates for monitoring termination. Regardless of the risk posed by a specific variable, it is important to establish baseline values for all variables before considering their removal. The removal of any locations or constituents from monitoring should be well documented and rationale should be provided for their removal.

Sampling frequency and resolution are determined by the end data use, as some models have specific data input requirements. Data may need to be collected semiannually for attributes such as aquatic life, while climate and water quality data may need to be measured continuously. Commonly accepted standards for sampling periods have been established for environmental sampling and monitoring, and usually consist of data collected over a minimum of one year to account for seasonal changes.

A sample list of considerations for baseline monitoring locations, variables and indicator organisms, and sampling periods is provided in Table 5.1. Categories and indicators to be monitored should be customized on a site-specific basis.

Impact Monitoring Design

Impact monitoring must be designed considering feasible data analysis techniques. Carefully choosing data analysis techniques and monitoring locations will produce a much more cost-effective program and give insight into sampling frequency and the duration of sampling. Impact monitoring

Table 5.1 Baseline monitoring design and considerations for mining operations (British Columbia Ministry of Environment, 2012; Leading Practice Sustainable Development Program for the Mining Industry, 2009b)

Baseline monitoring category	Indicators, measurements, and activities	Sampling period	Site selection
Geology and geochemistry	• Bulk sampling • Logging and sampling of drill cores • Background radiation • Mapping and sampling trenches • Regional and local surface mapping • Surficial geophysical studies • Stream sediment and surface water/ seep water surveys • Overburden mapping and sampling • Identification of surface features	—	• Must characterize geology and geochemistry over entire project footprint, especially areas where mineral extraction and drilling will take place.
Meteorology and climate	• Precipitation • Pan evaporation • Temperature (highs and lows) • Dew point • Duration of sunshine • Snowpack • Relative humidity • Barometric pressure • Sigma Theta (standard deviation of horizontal wind direction) • Wind speed and direction • Net radiation • Noise	One year of baseline data is usually necessary. Data should be collected daily or continuously.	• Many meteorological stations required if mine project extends into various topographical systems. • Must demonstrate weather and climate over entire project footprint.

Air quality	• Greenhouse gas emissions • Particulate matter ($PM_{2.5}$, PM_{10}) • Ozone • Nitrogen dioxide, sulfur dioxide • Volatile organic compounds (VOCs), PAHs • Heavy metals (Lead, arsenic, cadmium, nickel) • Carbon monoxide, carbon dioxide	One year of baseline data is usually necessary. Data should be collected daily or continuously.	• Many monitoring stations required if mine project extends into various topographical systems. • Must demonstrate air quality over entire project footprint.
Surface hydrology	• Stage (water level) • Flow and pump rates • Discharge flows • Water storage capacities • Site water balance • Photographs	Minimum of 2 years of flow data are required to evaluate accuracy of rating curves. Data should be collected daily or continuously.	• Based on physiographic features and mine plans. • Sites typically done on a case-by-case basis. • Multiple stations required in study area for redundancy and adequate characterization.
Hydrogeology	• Inventory of nearby wells and groundwater and surface water uses • Water levels in surface water • Pumping tests • Packer tests (including slug tests) • Fracture characterization • Borehole flow meters and geophysics • Groundwater monitoring wells • Exploration holes • Tracer tests	Groundwater data typically requires a minimum of 1 year of data. Data should be collected monthly or quarterly.	• Sites should be representative of different hydrostratigraphic units. • Monitoring well spacing depends on hydrogeological uniformity. • All permeable units affected by mining operation should be monitored. • Up-gradient monitoring should be done for background control.

Continued

Table 5.1 Baseline monitoring design and considerations for mining operations (British Columbia Ministry of Environment, 2012; Leading Practice Sustainable Development Program for the Mining Industry, 2009b)—cont'd

Baseline monitoring category		Indicators, measurements, and activities	Sampling period	Site selection
Terrestrial ecosystems		• Evaluate local or on-site streams • Evaluate natural discharges (springs, seeps, etc.) • Evaluate groundwater withdrawal, recharge, exchange, flow, resistance to flow, storage • Indicator fauna (e.g., mammals, avifauna, invertebrates, stygofauna, etc.) • Plant biodiversity • Vegetation coverage • Tree decline or dieback • Fire risk • Weeds and pests • Carrying capacity • Crop production	Data should be collected for a minimum of 1 year. Data should be collected annually or semiannually.	• Sites should be representative of different ecosystems of significance. • Areas should correspond with other sampling areas if possible. • Areas selected should have consistency before, during, and after mining operation.
Water quality	Ground water	• Dissolved oxygen • Flow • Odor • pH • Redox potential (ORP Eh) • Specific conductance • Hardness • Dissolved and suspended solids	Groundwater data typically requires a minimum of 1 year of data. Data should be collected monthly or quarterly.	• Directly down gradient of potential seepage areas (e.g., tailings impoundments). • Locations where impacts are possible at more than one depth require multi-leveled monitoring wells.

Surface water	• Temperature • Turbidity • Chemical oxygen demand (COD) • Biological oxygen demand (BOD) • Organics (dissolved organic carbon, total organic carbon) • Radionuclides • Tritium • Cyanide species • Nutrients (nitrate, nitrite, ammonia, total nitrogen, orthophosphate, total phosphate) • Anions (alkalinity, fluoride, chloride, bromide, sulfate) • Total and dissolved metals (aluminum, antimony, arsenic, barium, beryllium, bismuth, boron, cadmium, calcium, chromium, cobalt, copper, iron, lead, lithium, magnesium, manganese, mercury, molybdenum, nickel, potassium, selenium, silicon, silver, sodium, strontium, thallium, tin, titanium, uranium, vanadium, zinc) • Total phenol or BTEX (benzene, toluene, ethylbenzene, xylene) • Polycyclic aromatic hydrocarbons (PAH)	Surface water data typically requires a minimum of 1 year of data. Data should be collected on a monthly basis with weekly sampling during periods of maximum fluctuations and flows.	• Mineralized zones and ambient down-gradient areas. • Directly up-gradient from all potential mining impacts. • Use of existing water supply wells may be acceptable. • Upstream and downstream of discharge points, seepage points, and non-point contaminant sources. • Reference sites should be upstream of affected watershed, or in adjacent watersheds. • Multiple samples may be necessary for stratified lakes.

Continued

Table 5.1 Baseline monitoring design and considerations for mining operations (British Columbia Ministry of Environment, 2012; Leading Practice Sustainable Development Program for the Mining Industry, 2009b)—cont'd

Baseline monitoring category	Indicators, measurements, and activities	Sampling period	Site selection
Aquatic sediments	• Particle size distribution • Sediment transport rates • Total organic carbon • PAH • Metals (aluminum, antimony, arsenic, barium, beryllium, bismuth, boron, cadmium, calcium, chromium, cobalt, copper, iron, lead, lithium, magnesium, manganese, mercury, molybdenum, nickel, potassium, selenium, silicon, silver, sodium, strontium, thallium, tin, titanium, uranium, vanadium, zinc) • Moisture content • Sulfur	Sediment data is typically required once per year during low flow periods of late summer.	• Main stem depositional zones, lentic habitats, lotic habitats. • Sites should be upstream, downstream, and adjacent to the mining project.
Aquatic tissue residues	• Tissue samples from: Periphyton and macrophytes, benthic invertebrates, fish • Moisture content • Metals (aluminum, antimony, arsenic, barium, beryllium, bismuth, cadmium, calcium, chromium, cobalt, copper, iron, lead, lithium, magnesium, manganese, mercury, methyl mercury, molybdenum, nickel, potassium, selenium, silver, sodium, strontium, thallium, tin, titanium, uranium, vanadium, zinc) • Phosphorus	Tissue sampling is typically required once per year, usually during low flow periods.	• Largely based on where organisms are available (e.g., wetlands, oxbows, lakes, backwaters, streams, etc.). • Targeted areas should include affected water bodies and areas upstream, downstream, and adjacent to mining project.

Aquatic life	• Benthic macroinvertebrates • Periphyton	Benthic invertebrates and periphyton are collected once per year, usually from August to September. Two years are required for inter-annual variability.	• Areas should correspond with water, sediment, and tissue sampling. • Bodies of water selected should have consistency before, during, and after mining operation.
Fish and fish habitat	• Fish presence or absence • Fish species, life stages present • Indicators of fish abundance • Fish distribution • Life history timing (e.g., migrations) • Fish habitat (macrohabitats, mesohabitats, and microhabitats) • Fish production potential • Characterize riparian zones • Biogeoclimatic zones • Channel morphology	A minimum of 2 years of data collection prior to construction is recommended.	• Sites depend on viable fish habitats and can include spawning zones, holding pools, migration corridors, production areas, and areas where fish congregate. • May include all water bodies impacted by mining project • Reference sites should have similar hydrology, water chemistry, fish populations, and habitat characteristics.
Livelihood and socioeconomic	• Employment • Income • Compensation • Cost of living • Job creation • Economic diversity	Monitoring should be conducted as required or on a yearly basis.	• Local and regional communities

Continued

Table 5.1 Baseline monitoring design and considerations for mining operations (British Columbia Ministry of Environment, 2012; Leading Practice Sustainable Development Program for the Mining Industry, 2009b)—cont'd

Baseline monitoring category	Indicators, measurements, and activities	Sampling period	Site selection
Social	• Community networks • Social and civic participation • Local leadership • Social disturbances (e.g., noise, traffic, dust)	Monitoring should be conducted as required or on a yearly basis.	• Local and regional communities
Local communities	• Infrastructure • Buildings • Transportation • Community facilities • Contamination events	Monitoring should be conducted as required or on a yearly basis.	• Local and regional communities
Land use	• Economic value of land • Grazing, agriculture, horticulture, fishing, hunting • Recreation and natural aesthetics • Assets and natural resources • Cultural heritage and spiritual use • Archaeological sites	Monitoring should be conducted as required or on a yearly basis.	• Local and regional communities
Human	• Education • Immigration or migration • Health and health services • Exposure to hazardous materials • Safety	Monitoring should be conducted as required or on a yearly basis.	• Local and regional communities

must maintain effectiveness and should also meet the requirements for sound statistical analyses.

Taking all those considerations into account, a popular impact monitoring method used in mining is the Before-After-Control-Impact method (BACI). Before-after pertains to the time before and after a mining activity occurs, while control-impact pertains to areas assumed to not be impacted (control) and areas that are affected by the activity (impact). Impacts from mining activities can be expected or unexpected, so there needs to be monitoring at all times when an impact is a possibility. When choosing control and impact sites, the sites should be similar, but do not need to be identical. Measurements can be made from the change in the difference between the control site and the impact site.

Indicators

Indicators are the easiest and most cost-effective way to monitor health and ecosystem, economic, and social conditions. There is no way to feasibly monitor every part of the ecosystem; indicators must be chosen carefully in order to accurately represent the impacts of mining activities on the environment. Possible indicators for multiple environmental monitoring categories are provided in Table 5.2. Some criteria that can be used when choosing indicators include (Leading Practice Sustainable Development Program for the Mining Industry, 2009b):

- Validity – logically measuring factors they are intended to measure
- Reliability – remaining consistent over the course of the project
- Simplicity – not being overly complicated, particularly if the community is to participate in data collection
- Comprehensiveness – encompassing the whole complexity of project
- Availability – being easy to collect
- Practicality – not being excessively resource intensive

All of these considerations cannot be applied for every situation, due to budget and time constraints. Certain characteristics may be emphasized more heavily than others depending on the parameter being measured and the nature of the mining project. In other cases, it may be preferable to use information collected by local agencies, if available, to reduce costs. For mining projects that last multiple decades, choosing reliable indicators is especially important, as community and environmental monitoring needs may change over the course of the project.

Table 5.2 Impact monitoring design and considerations for mining operations (Leading Practice Sustainable Development Program for the Mining Industry, 2009b)

Impact monitoring category	Indicators, measurements, and activities	Sampling period
General	• Hydrocarbon consumption • Hazardous materials consumption • Greenhouse gas emissions (direct and indirect) • Energy consumption (direct and indirect) • Geotechnical hazards (collapse, landslides, subsidence) • General air and water quality parameters (see baseline measurements in Table 5.1) • Soil stockpile quantities, quality, and longevity	Monitored as required by permits.
Ore crushing and processing facilities	• General water quality parameters (see baseline measurements for water quality in Table 5.1)	Sampling should occur weekly to quarterly depending on the activity. Some monitoring should occur on an event-based basis.
Waste rock and ore stockpiles	• Waste rock and ore production rates • Volume and mass of stockpiles • Geochemical characterization • Erosion, subsidence, landslips • Infiltration rates • Moisture content in rock piles (pore volume, pore pressure) • Seepage flow pathways • Flow rates (surface runoff, surface seepage, leachate) • Geophysical surveys	Sampling should occur weekly to quarterly depending on the activity.

Tailings storage facilities and dams	• General water quality parameters (see baseline measurements for water quality in Table 5.1) • Radiation emissions • Milling and tailing production rates • Volume and mass in tailings facilities • Geochemical characterization • Tailings density and strength • Erosion, subsidence, landslips • Water volume, flow rates, decant rates, spillway flow rates • Infiltration rates • Seepage flow pathways and rates • Geophysical surveys	Sampling should occur weekly to quarterly depending on the activity.
Pits and open cuts	• General water quality parameters (see baseline measurements for water quality in Table 5.1) • Avifauna, mammals, and livestock • Mass and volume exposed to oxygen • Dewatering pump rates • Geochemical characterization • Groundwater levels and flow rates • General water quality parameters (see baseline measurements for water quality in Table 5.1) • Noise • Vibrations and blasting over-pressure • Erosion, subsidence, landslips	Sampling should occur weekly to quarterly depending on the activity. Some monitoring should occur on an event-based basis.

Continued

Table 5.2 Impact monitoring design and considerations for mining operations (Leading Practice Sustainable Development Program for the Mining Industry, 2009b)—cont'd

Impact monitoring category	Indicators, measurements, and activities	Sampling period
Underground mines	• Mass and volume exposed to oxygen • Dewatering pump rates • Geochemical characterization • Groundwater levels and flow rates • General water quality parameters (see baseline measurements for water quality Table 5.1) • Noise (above ground) • Vibrations and blasting over-pressure • Wallrock stability, subsidence, landslips • Underground air quality (see baseline measurements for air quality in Table 5.1)	Sampling should occur weekly to quarterly depending on the activity. Some monitoring should occur on an event-based basis. Air quality should be monitored continuously.
Heap and dump leach piles	• Ore production rates, mass, and volume • Geochemical characterization • Surface water levels and flow rates • Groundwater levels and flow rates • General water quality parameters (see baseline measurements for water quality Table 5.1) • Erosion, subsidence, landslips	Sampling should occur weekly to quarterly depending on the activity. Some monitoring should occur on an event-based basis.
Roads and drill pads	• Erosion, subsidence, landslips • General water quality parameters (see baseline measurements for water quality Table 5.1) • Noise • Moisture content • Organism passage	Sampling should occur weekly to quarterly depending on the activity. Some monitoring should occur on an event-based basis.

Quarries	• General water quality parameters (see baseline measurements for water quality Table 5.1) • Noise • Vibrations and blasting over-pressure • Erosion, subsidence, landslips	Sampling should occur weekly to quarterly depending on the activity. Some monitoring should occur on an event-based basis.
Smelting, power generation, etc.	• Stack and flue emission air quality (see baseline measurements for air quality in Table 5.1)	Air quality should be monitored as required by permits.
Water storage, sediments basins, reservoirs	• Flow rates • Water levels • Storage capacity, residence time, sediment analysis • General water quality parameters (see baseline measurements for water quality Table 5.1) • Erosion, subsidence, landslips	Sampling should occur weekly to quarterly depending on the activity. Some monitoring should occur on an event-based basis.

Baseline monitoring should continue throughout the course of the mining operation and is not included in this table.

Sampling Considerations

There are many factors to consider when choosing appropriate instrumentation or sampling procedures for monitoring. Many of the general criteria used when choosing an indicator can also be used for choosing instrumentation and sampling techniques (e.g., practicality, validity, reliability, etc.). However, there are four additional factors that should be considered for sampling:

- Real-time data collection and processing
- Detection limits
- Sampling sites and methods
- Quality assurance and quality control protocols

Real-Time Monitoring

Real-time monitoring is when instruments can record observations and process the data at that time. The data can be sent over a network to a computer that outputs the data for analysis and interpretation. Real-time monitoring can be useful in remote locations to reduce frequency of visits, resulting in increased safety benefits and lower costs while providing data at a greater frequency than conventional methods. The real-time access to information can allow operators to act proactively, rather than reactively, whereby reducing non-compliance events and reducing fines and cleanup costs. Real-time monitoring can also be helpful if alarms are needed when compliance limits or "trigger values" are exceeded.

When choosing real-time instrumentation, the following factors need to be considered:

- *Telemetry bandwidth available*: Is there adequate bandwidth to support the quantity of data that needs to be relayed over the network, and will there be adequate bandwidth if future expansion is required.
- *Network support*: Is there adequate personnel to maintain network operations and maintain continual operation?
- *Existing network infrastructure*: Is there already existing infrastructure (e.g., radio telemetry, mobile telephone system, satellite telemetry, etc.) that can be used to implement a real-time system?
- *Communications protocol*: Do the field instruments use a communication protocol that can be accepted by the telemetry units?
- *Power consumption*: How will these remote devices be powered and how long will each charge last? Is enough power available to send data in the required quantities?

- *Maintenance requirements*: How often does the equipment need to be serviced or replaced? How do the instruments perform under non-ideal conditions?
- *Data delivery, storage, and display*: Does the device have enough memory for required storage and delivery?
- *Geographical coverage*: What coverage is required (can signals be transmitted over this distance?) and will vegetation or topography attenuate signals?

Real-time monitoring may not always be a viable option for all mining projects, depending on the limitations of the telemetry platforms available and local conditions.

Detection Limits

A detection limit is the lowest amount of chemical (or water quality parameter) an instrument can detect in solution. When considering instrumentation detection limits or various sampling procedures, it is important to choose methods with the smallest detection limit available (within reason) as environmental policy and analytical methods are continuously pushing for lower and lower detection limits. As technology and policies change, a high detection limit will result in data being obsolete. For this reason, data that needs to have long-term relevancy, such as baseline data, should have a more rigorous detection limit. Baseline monitoring data should thus use cutting edge instrumentation that offers low detection limits. However, other impact monitoring data might not need such a rigorous detection limits and thus mid-priced commercial analysis might be acceptable.

It is important to realize that analyses with very low detection limits may also require rigorous and extensive sampling techniques and preparation work that may not always be feasible. Each parameter will need to be evaluated to determine which detection limit is the most practical (Leading Practice Sustainable Development Program for the Mining Industry, 2009b). All chosen methods and associated detection limits should be thoroughly documented with the collected data.

Selection of Sampling Sites and Methods

When selecting sampling and monitoring sites, the following should be taken into account:

- Activities that will contribute to contamination or the spread of contamination

- Areas and locations where contamination is most likely to occur
- Areas and locations where contamination events have the greatest likely impact on environmental attributes
- Heterogeneity within the sampling location (distributional, compositional, or morphological)
- Representativeness of the location, in relation to the project area
- Cost, accessibility, safety, and reliability
- Site disturbances caused by sampling and monitoring activities

For example, common water quality sampling locations may include point discharges, downstream of a confluence, groundwater up-gradient and down-gradient of the mining operation, and nearby surface water resources.

Instrument selection and sample collection will depend on what type of sampling method will be utilized. Sampling methods can include the following:

- Grab samples
  - Samples taken that represent a snapshot in time, typically collected by hand, although instruments are available
  - Typically used for homogeneous materials like water
  - May require transverse and longitudinal surveys to determine representative locations for sampling
  - May have to take multiple samples to account for stratification in bodies of water
- Semi-continuous and continuous sampling
  - Automated instruments and sensors are used to collect samples for a given time interval specified by the needs in the monitoring plan
  - Sampling may be triggered automatically or manually based on various ambient conditions (e.g., rainfall, temperature, flow)
  - Care must be taken in sampler placement to ensure debris do not prevent sample collection
  - Samples may be many discrete samples, or a composite sample
  - Samples and data can be stored and collected during maintenance activities
- Passive sampling
  - Devices that can be deployed in the field for a period of time, after which they can be collected and analyzed for a time averaged value
  - These systems are often low cost and can be deployed in large numbers to increase coverage
  - Passive sampling typically results in a composite sample

- Remote surveillance
  - Instruments are used to collect data and communicate that data back to a station for analysis
  - Equipment can be very discrete or hidden to reduce vandalism or theft
  - Monitoring programs can include alarms when concentrations exceed limits specified by the monitoring plan
- Remote sensing
  - Satellites or aircraft can be used to monitor changes in reflected or emitted radiation, though typically sunlight
  - Can collect data over an extremely large area
  - Can monitor areas that may otherwise be unreachable

Different parameters will require different types of sampling methods. Sampling methods may be chosen based on site accessibility, site disturbance, frequency of sampling required, cost, potential hazards, and analysis techniques. It is important that samples be taken at roughly the same time period to reduce time dependent variations (e.g., diel or tidal variations).

Quality Assurance and Quality Control

A quality assurance and quality control (QA/QC) program should be established to ensure the collection of credible data and to maintain all relevant QA/QC procedures and guidelines. Common QA/QC documentation should include (British Columbia Ministry of Environment, 2012; Environment Canada, 2011):

- Sampling procedures and guidelines
- Sampling storage, handling, processing, and chain of custody
- Laboratory procedures, analyses, sampling handling, and calibration
- Operational limits for sensors
- Sensor drift and calibration procedures
- Error bounds on sensors, data loggers, etc.
- Sensitivity analyses
- Records of field visits, maintenance, and calibration
- Special calibration or corrections required for temperature or pressure
- Protocols for incorporating and evaluating new analytical methods or technologies
- Support for sampling and analysis development when necessary

When collecting field samples, special care should be taken to use specified sample bottles, containers, and apparatuses, as well as follow methods for sample handling, holding times, specific preservation methods, storage temperatures and conditions, and chain of custody. Sampling technique should be specified and samples should be taken at regular intervals at the same location, taking note of any observations or special conditions that may have affected the sample. For example, groundwater wells may require purging, especially if they are not normally used as a water source, before indicator parameters stabilize. At least three successive readings should be taken to ensure all readings have stabilized. All employees must be thoroughly trained in proper sampling techniques and data reporting to maintain consistency.

Duplicate samples should be taken at randomly selected locations to ensure reproducibility of the sampling methods. For highly variable samples, such as sediment samples, field splitting can also be used to check for variability. Increasing the sample size can also reduce sampling error when dealing with samples that exhibit high heterogeneity, low constituent concentration, large particle sizes, or when a higher degree of confidence is desired. Sample blanks, or samples that consist of purified water, air, or another known substance, should be taken into the field and stored under the same conditions as field samples to determine the effects (if any) of sampling handling, bottling, and storage on the sample properties.

Proper maintenance of all monitoring and sampling equipment is essential for the success of a monitoring program. Sampling and monitoring equipment should be calibrated against known standards prior to every sampling trip or as specified in the QA/QC plan. Continuous monitoring equipment should be enclosed in a weatherproof housing to protect sensitive electronics and prevent loss of data. The portion of the equipment that is taking measurements is, by its very nature, exposed to the environment and should be thoroughly cleaned and calibrated on a regular basis. Sensor calibration for continuously monitored stations should be checked, at a minimum, quarterly for sensor drift. Manual readings should be taken at each site and compared to the station readings to ensure accurate measurements. Sensors should be recalibrated once sensor drift exceeds 2% (British Columbia Ministry of Environment, 2012). Monitoring stations should be outfitted with solar panels or battery backups to prevent data loss and continuously logged data should be downloaded at regular intervals to prevent data loss from malfunctions.

Laboratory analysis for specific constituents requires additional QA/QC procedures. Laboratories are encouraged to use standard methods and seek accreditation such as ISO 17025 standard (British Columbia Ministry of Environment, 2012). Laboratory analyses often require analyzing blank samples and laboratory duplicate samples to ensure reproducibility of the analyses. Other samples, such as chemical standards, calibration curve samples, matrix spikes, and laboratory control samples, should also be run to ensure lab equipment is properly calibrated throughout an analysis and to detect possible interferences. Independent, accredited outside laboratories should run a quality assurance audit program to ensure accurate laboratory results. A robust QA/QC program will ensure compatibility and comparability between analyses performed at different laboratories, and has a protocol for rectifying discrepancies in reported data (Environment Canada, 2011).

Adaptability and Longevity

As mentioned above, the implemented monitoring plan needs to be adaptable and relevant over the long term. Adaptive management programs can be integrated into monitoring plans, allowing for periodical evaluation and improvements as situations change and environmental knowledge evolves (Government of Alberta, 2012). In order to achieve this goal, various monitoring plans for different time spans need to be created. Knowing long-term goals will assist in the development of details for short-term goals.

In short-term annual monitoring plans, procedures and objectives should be clearly defined and documented. Key elements such as responsibilities for data management, storage, interpretation, final reporting, and recommendations need to be defined. In addition, a project manager needs to be assigned to coordinate monitoring tasks and make sure that all activities are implemented correctly. Short-term plans should maintain a set of alternative sampling locations, methods, and equipment to account for changing environmental conditions and unforeseeable situations.

In medium-term monitoring plans of 3–5 years, the monitoring program should be linked to the medium-term construction projects and production plans. For example, if the mine needs to expand their operational area, pre-clearing monitoring will then be needed in that area even if it was not previously required.

The long-term monitoring plan of the life-of-mine needs to be reviewed periodically based on the rate of change of mining activities. For abandoned mines or mines with suspended operations, it is very beneficial to have a record of past monitoring operations, data, analyses, and reports.

It is vital to maintain accurate data reporting and maintain all associated databases throughout the life of the project. Data and metadata must be thoroughly documented, as it is likely that those who initially collected the data will not be the same people who will analyze it due to employee turnover. The long length of mining projects can mean that data collected near the start of the project may be stored in formats that become outdated or inaccessible by the end of the project. Implementing proper database maintenance and migrating to new software when appropriate can prevent accessibility issues.

Specific Monitoring Plans

Because each environmental attribute can be measured in a variety of ways and may require unique quality objectives, each attribute needs a specific monitoring plan. Most mining operations have, at a minimum, the following monitoring plans: air quality, water quality, geotechnical and geochemical, and ecological. Monitoring plans should contain procedures for both baseline monitoring and impact monitoring.

Air Quality

There are many sources of emissions at a mining site that should be monitored. Emission sources may include stack and flue gas from smelting and refineries, vehicular emissions, and vehicular traffic and wind erosion. Some general monitoring considerations for air quality are listed as follows (Leading Practice Sustainable Development Program for the Mining Industry, 2009a):

- Baseline monitoring should measure indicators specified by regulatory requirements and the risk assessment; measurements include particulate matter, heavy metals, ozone, sulfur dioxide, nitrogen dioxide, and methane
- Instrumentation should measure continuously collected data averaged over different time intervals, depending on the air emission being measured and weather conditions
- Weather stations are required for collecting meteorological data to incorporate into analysis
- Dust gauges and other passive dust collection methods for deposition monitoring can monitor dust emissions to reduce community dissatisfaction
- Specific target compound should be monitored if the risk assessment indicates that it might present a problem (e.g., radionuclides, silica, etc.)
- Specific locations for monitoring should be determined based on weather and wind conditions

- Data from weather stations, satellite imagery, remote sensing, and air quality models should be integrated to assess trends, pathways, and contaminant concentrations
- Air quality monitoring upwind and downwind of the mining operation can help determine negative impacts

Water Quality

There are many possible sources of water contamination at a mining site that should be monitored. Possible sources may include acid mine drainage, wastewater discharge, storm water discharge, leachate, tailings ponds, and water storage facilities. Some general monitoring considerations for water quality are listed as follows (Leading Practice Sustainable Development Program for the Mining Industry, 2009a):

- Recommended monitoring locations:
  - Where there are large fluxes of water in and out of operation
  - Where water quality is significantly altered
  - Where acid mine drainage is likely to occur
  - Where an operational task is sensitive to changes in water quality
  - If a hazard is posted to safety or human and ecosystem health
  - Upstream or up-gradient and downstream or down-gradient of the mining operation
- Baseline monitoring should include reference catchments: areas that are not affected by mining activities, but are similar in other features (e.g., size, meteorological conditions, topography, and ecological conditions) so that natural variability in the ecosystem can be taken into account
- On-site monitoring might need to be real time to support operator decisions
- On-site monitoring should evaluate discharges, storage, holding dams, and groundwater
- Receiving systems and reference sites should be monitored off-site to provide evaluation of mining impacts, seasonal variability, and impacts of non-mining activity
- Chemical monitoring should be integrated with biological monitoring

Geotechnical and Geochemical Conditions

Geotechnical and geochemical conditions need to be monitored to ensure the stability of the mining site and to characterize possible risks from exposed waste rock and tailings. Some general considerations for

geotechnical and geochemical monitoring are listed as follows (Leading Practice Sustainable Development Program for the Mining Industry, 2009a):
- Monitor mining site for erosion, subsidence, landslides, and rock falls
- Geotechnical stability and safety should be monitored through daily inspections, radar scanning, and survey prisms
- Determination of geochemical properties of ore, waste rock, and surrounding area can be used to evaluate risk of acid mine drainage or heavy metal contamination
- Monitor geochemical characteristics and update models if necessary

Ecological Conditions

Ecological conditions need to be monitored to ensure the negative impacts on flora and fauna are kept to a minimum. Some general considerations for ecological monitoring are listed as follows (Leading Practice Sustainable Development Program for the Mining Industry, 2009a):
- The scale of ecological monitoring to be implemented is based on the estimated scale of impacts
- Baseline conditions should be established that appropriately capture seasonal variation; monitoring should span a number of years to take into account variations across different years.
- Document types of species present, where they occur, abundancy, soils types, vegetation types, and changes over time
- Evaluate patterns between species and different threatening processes for:
  - Flora and vegetation
  - Vertebrate fauna
  - Land-based invertebrate fauna
  - Aquatic invertebrate fauna
- Monitor during operations to measure the extent of mining activity impacts
- Monitor during rehabilitation to measure whether rehabilitation objectives are met

5.3 DATA MANAGEMENT

Assuring quality data are collected and analyzed is crucial to the success of monitoring programs. Rigorous data entry QA/QC procedures are necessary to ensure data are accessible, accurate, and protected from tampering or unauthorized users. The following suggestions provide some security measures (Leading Practice Sustainable Development Program for the Mining Industry, 2009b):

- Require authorization for database access
- Provide tracking of edits
- Spot check data entry and analysis for errors
- Back up databases both electronically and via hard copies and regularly check integrity and quality of backups
- Store raw data files separately from databases in the event of corruption

In addition to quality assurance measures, it is important to make sure that the means of collecting data and reading data are clear and the knowledge of how to use the databases are easily transferred. Carefully documented protocols, such as standard operating procedures, should be developed for monitoring techniques, data collection and entry, and sampling procedures and location selection. Procedures, reports, and protocols should be recorded in such a manner that new personnel are able to continue the monitoring programs when old employees leave. A robust spatial database such as a geographic information system (GIS) is a valuable tool for keeping track of monitoring locations and sampling information.

Due to the continually evolving nature of information technology, software used to record data must be standardized or readily transferable to other software to avoid obsolescence. Helpful software features that allow the user to set up an automated data quality checks, can provide data quality scores to be associated with stored measurements (Leading Practice Sustainable Development Program for the Mining Industry, 2009b).

Depending on the nature of the project, monitoring data may need to be organized and uploaded onto the Internet to ensure transparency and free access for concerned parties. This would enable independent organizations to analyze the data and draw their own conclusions on environmental impacts. Such publically available data should include relevant information regarding standard operating procedures and QA/QC protocols (Environment Canada, 2011). It is vital for operations that make data publically available do so in a timely manner, as delaying data publishing may lower stakeholder confidence and arouse suspicion from environmental groups.

Data Analysis Techniques

Monitoring data should be analyzed quickly after collection to enable modification of monitoring practice, if necessary, and quick feedback to stakeholders and operators. Samples should represent naturally occurring events, so if analysis shows that data collection does not represent conditions as they occur, monitoring needs to be modified. For instance, more stringent "trigger" levels to alert managers to a problem can allow more in depth investigation into the impacts of certain activities. To facilitate the analysis process, recording observations may help data interpretation. Observations can include such factors such as dead fish, algal blooms, yellowing trees, floods, and any issues with monitoring equipment. However, since observations can be subjective, it is important to be as quantitative as possible and to word observations with caution.

The simplest data analysis should compare and contrast both quantitative and qualitative (observational) data from different areas in the monitored area. Data should also be compared to baseline levels and other control locations previously selected. In some cases, such as groundwater monitoring, individual data points may not be as informative as median or long-range values for an entire region. Graphical and tabular data are often preferable to allow for easy visualization of differences and trends. When possible, conceptual models should be developed to visualize parameters, such as groundwater divides, boundaries, and hydrostratigraphy, in order to provide context for the collected data (British Columbia Ministry of Environment, 2012).

Data analyses should include both statistical analyses as well as an in depth analysis that details environmental mechanisms of action for the stressors. In the statistical analyses, parametric analysis is preferred over non-parametric analyses (Leading Practice Sustainable Development Program for the Mining Industry, 2009b). However, because environmental sampling generally has large variability and small sample sizes, non-parametric or Bayesian statistical tests may be used on non-normal data. If environmental modeling is used, calibration and validation statistics, as well as model assumptions should be clearly stated.

Levels of uncertainty, anomalous data, precision, accuracy, comparability, and the completeness of data should always be reported with analysis results to provide a complete picture to decision makers (British Columbia Ministry of Environment, 2012). At a minimum, data reporting should include frequency of analysis, variability of measurements, and any methods used for data transformation, such as weighting, change-point analysis, and selection criteria. Furthermore, rationales for using the selected analytical methods should be provided.

5.4 MONITORING TECHNOLOGIES

Air Quality Monitoring

Emissions monitoring has become increasingly important to meet emissions limits or guidelines and for maintaining accurate emissions inventory data. Emissions are typically monitored using either a continuous emissions monitoring system (CEMS) or a parametric monitoring system (U.S. Environmental Protection Agency (U.S. EPA), 2002). Continuous emissions monitoring systems measure pollutants from a sample taken directly from a stack, duct, or emission point. Continuous emissions monitoring systems consist of a sampling and condition system, the gas analyzer, and data acquisition systems. Emissions can be monitored by placing the analyzer directly within the duct or stack (in–situ) or by extracting a sample via a sample probe for transport to the analyzer (extractive) (U.S. Environmental Protection Agency (U.S. EPA), 2002). Extractive methods allow for sample conditioning and protect against high temperatures, velocities, and pressure. In–situ measurement methods can be either path measurements, where a signal is sent across the stack and reflected back to a detector, or point measurements, where a sample cell is suspended in the emission stream. Continuous emissions monitoring systems can analyze a variety of parameters including SO_2, CO, O_2, NO_x, CO_2, particulate matter, VOCs, flow rates, and opacity (U.S. Environmental Protection Agency (U.S. EPA), 2002). Particulate matter measurements in CEMS usually rely on surrogate measurements using optical or electrostatic techniques and may not be accurate for highly variable emission streams (U.S. Environmental Protection Agency (U.S. EPA), 2002).

Parametric monitoring systems do not measure emissions directly, but instead measure key variables related to emissions, such as operational temperatures, pressures, and flows (U.S. Environmental Protection Agency (U.S. EPA), 2002). Parametric systems need to be carefully calibrated to ensure a reasonable correlation between measured parameters and emissions. For example, operational temperatures are strongly correlated to emissions of VOCs and pressure drop across a PM filter unit or SO_2 scrubber can indicate performance (U.S. Environmental Protection Agency (U.S. EPA), 2002). Parametric monitoring systems are used in conjunction with a variety of control technologies for PM, VOCs, NO_x, CO, and SO_2, and are commonly used for smaller emission sources.

Particulate Matter Monitoring

The monitoring of dust emissions is important for determining the effectiveness of preventative measures and for determining the potential hazard. Emissions monitoring systems like CEMS can monitor PM indirectly, but only if emission streams are constant and the system is properly calibrated (U.S. Environmental Protection Agency (U.S. EPA), 2002). For highly variable emissions, PM can be monitored using passive or active systems. Passive systems simply measure PM deposition over a period of time and involves collecting dust using a flat surface, glass slides, stick pads, bowls, or a cylindrical container (Department of Environment Climate Change and Water NSW, 2010). Passive systems are usually exposed and measured on a weekly or monthly basis. Active systems utilize gravimetric measurements to provide a time-weighted average PM concentration and are the standard EPA method for measuring PM (U.S. Environmental Protection Agency (U.S. EPA), 2002). In an active system, a measured volume of air is drawn through a filter and the difference in the weight of the filter is used to determine PM concentration (Department of Environment Climate Change and Water NSW, 2010; U.S. Environmental Protection Agency (U.S. EPA), 2002). A combination of both monitoring systems should be used to evaluate the effectiveness of dust control measures.

Personal and mobile dust monitors are available to measure individual exposure or to attach to vehicles to quantify dust generation during operation. These dust monitors can be active systems (utilizing gravimetric measurements), or may use light scattering as a passive method to determine dust exposure.

Methane Monitoring

Methane monitoring in underground mines is important to prevent catastrophic explosions, mine fires, loss of human life, and the venting of a potent greenhouse gas. Methane detectors typically use catalytic heat of combustion sensors or infrared sensors to detect methane (Kissell, 2006). Catalytic heat of combustion detectors are limited to concentrations of methane below 8% and oxygen above 10%. Infrared detectors measure the absorption of infrared light due to the presence of methane and while they can measure methane without the presence of oxygen, they are hindered by the presence of water vapor and dust (Kissell, 2006). Methane detectors are classified as either portable methane detectors or machine-mounted monitors. Portable methane detectors can be hand-carried to measure methane levels throughout the mine. Machine-mounted monitors are integrated with machine operation, where machines cannot operate without a functioning methane monitoring system to ensure safe operating conditions. Both continuous methane monitoring with machine-mounted detectors and intermittent monitoring (every 20 min) with portable detectors are required by law in many countries (Kissell, 2006).

Water Quantity and Quality Monitoring

Water Quantity Monitoring

Water quantity monitoring is important for maintaining adequate flows and water levels in aquatic habitats and for accurate mine water balances. Surface water quantity monitoring primarily involves measuring water level and flow rates. Water level, or stage height, is commonly measured using a fixed staff gauge, consisting of a graduated metal plate that is partially submerged and secured at a reference height (usually the streambed), from which the height of the water surface can be measured (Rantz, 1982; U.S. Environmental Protection Agency (U.S. EPA), 2001). Flow can be measured using a variety of methods, but typically involves using a weir or measuring the water velocity over the cross-sectional area of the stream (Rantz, 1982). Weirs are fixed barriers across a river or stream that force water to flow over their tops, where the height of the water above the weir can be used to calculate flow. In shallow and crossable bodies of water, water velocity is measured using a current meter, which is a simple device that is attached to a wading rod which measures water velocity through mechanical means (e.g., a rotor or an anemometer) or using an Acoustic Doppler Velocimeter, which determines flow by sending out sound waves and measuring the change in frequency observed in the return signal (Rantz, 1982; U.S. Environmental Protection Agency (U.S. EPA), 2001; U.S. Geological Survey (USGS), 2014). Other instrumentation for measuring velocity include electromagnetic and optical strobe meters (U.S. Environmental Protection Agency (U.S. EPA), 2001). In deeper bodies of water, velocity can be measured from a boat using an Acoustic Doppler Current Profiler, which sends sound waves to the bottom of the water body and operates in a manner similar to the Acoustic Doppler Velocimeter (U.S. Geological Survey (USGS), 2014).

Groundwater levels are important for determining flow directions and changes in gradients. Groundwater levels can be measured at monitoring or supply wells. There is a variety of equipment that can be used for monitoring water levels including chalked tape, weighted tape, pressure transducers, acoustic probes, floats, and pressure gages (Holmes et al., 2001). Chalked tape can be manually lowered into the well and water level measured according to the wetted portion of the tape measure. Pressure transducers can be lowered into the well to measure the height of the water above the transducer. Acoustic indicators measure the surface of the water by timing how long it takes for sound waves to bounce back from the surface of the water. Floating recorders can be used for continuous water level monitoring, where the amount of deployed cable attached to the float can be automatically logged. In flowing wells, a pressure gage can be attached to the well casing to determine the head above the measuring point. In addition to monitoring wells, piezometers can be installed to measure hydraulic head.

Water Quality Monitoring

Water quality monitoring is vital to determine any negative impacts mining operations have on local and regional surface and groundwater resources. Monitoring of water quality can be done by manually collecting water samples and measuring constituent values in the field or in a laboratory. Collecting surface water samples, at the most basic level, involves using a pole, submersible device, or one's hand to manually collect water in a bottle or container. Collection containers can be made of materials such as stainless steel, glass, polyvinyl chloride (PVC), silicone, or polypropylene, depending on the target analyses and possible interactions between the water and the container (Lane et al., 2003). Sampling equipment is generally classified as isokinetic depth-integrating samplers or non-isokinetic samplers. Non-isokinetic samplers are devices for which a water sample enters the device at different velocity than the ambient water velocity (Lane et al., 2003). Non-isokinetic samplers include open mouthed bottle samplers (e.g., handheld bottles or pole mounted bottles), weighted bottle samplers, BOD and VOC samplers, and thief samplers (e.g., Kemmerer samplers, Van Dorn samplers, and double-check valve bailers) (Lane et al., 2003). Isokinetic depth-integrating samplers continuously collect a representative sample where water enters the device at the same velocity as the ambient water (Lane et al., 2003). These samplers consist of a bottle or bag with an attached cap and nozzle to control water collection, and are either handheld or attached to a cable-and-reel system (Lane et al., 2003).

Some automated systems continually collect water samples over a period of time and aggregate them into one sample, which can be collected and analyzed as a point-integrated sample (Lane et al., 2003). Such devices are often employed at remote sites and ephemeral streams. Automated systems are commercially available from a variety of vendors including Teledyne ISCO, Campbell Scientific, and Xylem.

Groundwater samples are collected using pumps designed for monitoring wells, bailers, thief samplers, or from pumps installed on supply wells. Monitoring well pumps can be either suction-lift pumps (e.g., peristaltic pump) or submersible pumps (e.g., bladder pump, centrifugal impeller pump, rotor pump, progressive cavity pump) (Lane et al., 2003). Submersible pumps are preferred over suction-lift pumps because they do not create a vacuum, which can lower levels of dissolved gases and VOCs (Lane et al., 2003). Thief samplers and bailers, while generally not recommended for sampling as they disturb the water column, are useful for sampling very deep wells or groundwater with high concentrations of contaminants that can damage pumps (Lane et al., 2003).

When taking field measurements, simple handheld multi-parameter meters can be used to quickly determine major water quality parameters including pH, specific conductance, temperature, turbidity, and sometimes dissolved oxygen. These handheld meters are commercially available from a variety of vendors including YSI, Thermo Scientific, and Hach. Depending on accuracy and regulatory requirements, some parameters can be measured by relatively simple and inexpensive methods, as compared to handheld meters or deployable sensors. For example, salinity can be measured using a hydrometer or a refractometer, which use sample density and refraction, respectively, to determine salinity; surface water turbidity can be measured using a Secchi disk; pH and chloride levels can be measured using disposable test strips; temperature can be measured using a simple thermometer. To determine the concentration of other constituents of concern, such as heavy metals, water samples must be taken to a qualified laboratory for in-depth analysis.

Automated systems continuously monitor water quality over a period of time. Continuously monitoring systems rely on a series of sensors that are always on and continually log data concerning water quality parameters. The most common types of water quality sensors are temperature, specific conductance, pH, dissolved oxygen, and turbidity. Temperature is typically monitored using thermistors, semiconductors that change in resistance with temperature and are very reliable, durable, and accurate to $\pm1°C$ (Wagner et al., 2000). Dissolved oxygen can be measured using temperature compensated polarographic membrane sensors, which consume dissolve oxygen during measurements and require a constant flow for accurate measurements (Wagner et al., 2000). Specific conductance is measured using electrode based contact sensors that usually incorporate automatic temperature compensation (Wagner et al., 2000). Turbidity is measured using a light sensor that measures the light scattered by constituents in the water from a light emitting diode (Wagner et al., 2000). pH is commonly measured using a hydrogen–ion electrode, which measures a potential gradient across the glass membrane of the probe (Wagner et al., 2000). Multi-parameter continuous water quality monitoring probes such as Sondes and CTDs (conductivity, temperature, depth) are commercially available and commonly used throughout industry.

Continuous monitoring systems can be outfitted to wirelessly transmit data to a central location or may require periodic data collection from on-board storage. In urbanized areas, systems can transfer data directly using WiFi or cellular signals; however, in remote locations data may need to be transmitted via satellite. If remote transmission is not feasible, data must be logged and stored onsite. Data loggers for water quality monitoring are commercially available from a variety of vendors including Onset HOBO Data Loggers, Campbell Scientific, Geotech Environmental Equipment, and Xylem.

Land Impact Monitoring
Soil Monitoring
Soils in and around the potential mining site should be monitored for contamination and soil quality indicators, such as organic carbon, water-holding capacity, structure, rooting depth, pH, and nutrient availability. Beyond basic field observations such as soil structure, consistency, and color, most major soil constituents and nutrient analyses require soil samples be collected and analyzed in a qualified laboratory.

Topsoil can be sampled using basic tools such as shovels, scoops, and spades (U.S. Environmental Protection Agency (U.S. EPA) Region 9, 1999). Care must be taken to avoid using brass- or chrome-plated tools, which may contaminate soil samples; stainless steel equipment is commonly used. Square sampling templates may be used to ensure consistent sampling areas and distribution (Carter and Gregorich, 2008). Shallow subsurface soil samples may require an auger to bore a hole from which a sample can be collected using a tube core sampler (U.S. Environmental Protection Agency (U.S. EPA) Region 9, 1999). Bucket augers can be used to directly recover a sample, as compared to posthole augers, which are meant primarily for drilling into the soil. For deep subsurface soil samples, vehicle mounted hydraulic probes and coring devices are necessary (U.S. Environmental Protection Agency (U.S. EPA) Region 4, 2014; U.S. Environmental Protection Agency (U.S. EPA) Region 9, 1999). Coring methods depend on the nature of the soil or rock being sampled and include direct push coring, hand coring, and mechanical operations such as vibracoring, percussion sidewall coring, and rotary sidewall coring. Various coring devices can be utilized including dual tube soil samplers, standard or split spoon samplers, and Shelby tube samplers (U.S. Environmental Protection Agency (U.S. EPA) Region 4, 2014). Soil and mineral core samples collected during preliminary exploration activities may assist in soil characterization.

Slope Stability Monitoring

Unstable slopes in open-pit mining are a serious hazard and slope failures can cause many deaths. The primary factors governing slope stability are the geometry, geology, physicomechanical properties of the rock and soil, groundwater hydrology, and other factors including the effects of earthquakes (He et al., 2008). Slopes should be monitored for warning signs of instability, such as tension cracks, abnormal water flow, bulges, and rubble. Monitoring for such signs can involve simple solutions such as comparing changes in pictures over time, or using tape, paint, or flags to mark the ends of cracks to easily visualize any new crack propagation (Girard, 2001). Wireline extensometers, consisting of a wire attached to an unstable area with the other end attached to a pulley and weight on stable ground, can be used to visualize slope movement by the height of the weight (Girard, 2001). Surveying networks consisting of target prisms located at areas of possible instability and a fixed control point for surveying can be used to determine trends in slope movement (Girard, 2001).

Subsurface slope monitoring can be done using borehole probes, inclinometers, and time-domain reflectometry. Borehole probes are extremely simple and consist of a metal probe lowered down a borehole, where any shifts or movement along a slide plane that intersect the hole will cause the probe to become stuck at that intersection (Wyllie and Mah, 2004). Inclinometers are probes that contain accelerometers used to measure the tilt of the probe. When lowered down a borehole, inclinometers can provide an accurate reading of shifts within the borehole caused by slope movement (Wyllie and Mah, 2004). Time-domain reflectometry can locate a sliding surface and monitor movement by grouting a co-axial cable into a borehole, where any crimps or kinks in the cable caused by slope movement can be detected by changes in the impedance of the cable (Wyllie and Mah, 2004). Advanced technologies such as the embedment of rock bolts in the potentially unstable mining slopes and monitoring of strains in the bolts can provide an indication of crustal movement of slopes that may not be visible otherwise.

Subsurface monitoring of stress within the slope can provide an earlier indication of slope instability as compared to surface monitoring due to continual changes in stress prior to slope displacement (He, 2009). Using available information on slope geology, geometry, and surface and sub-surface activity, 3D numerical modeling software, such as FLAC3D, can be utilized to evaluate slope design and stability (He et al., 2008).

Subsidence Monitoring

During the operation of a mine, subsidence should be monitored daily, weekly, or based on certain events such as seismic activity (Leading Practice Sustainable Development Program for the Mining Industry, 2009b). Monitoring of subsidence can be done in several ways. Monitoring is historically done through ground surveying. This technology uses automatic digital leveling, total stations (electronic distance measurement), and GPS receivers (Ge et al., 2007). The monitoring is constrained to localized areas and does not monitor any regional deformation. Another technology used to monitor subsidence is differential interferometric synthetic aperture radar (DInSAR). This technology can take multiple observations using a high resolution to measure land deformation with a high degree of accuracy. Due to errors from atmospheric conditions when using radar, DInSAR technology should be combined with GPS and GIS technologies (Ge et al., 2007).

5.5 EMERGING MONITORING TECHNOLOGIES

Many advances have been made in the field of monitoring technologies in the past several decades. Technology improvements allow more thorough and efficient data collection, make sites accessible that were previously difficult or dangerous to access, lower detection limits, allow collection of data from species that were not previously monitored, lower costs, and increase the ease of data analysis. Areas of notable improvement include sensor technology, remote sensing methods, and non-destructive animal sampling and tracking.

- Improvements in sensor technologies
  - Biosensors: Accurate and sensitive means of rapid and early detection of pollution in water (Long et al., 2013)
  - Geosensor networks: Assess plant growth and land cover, real-time event detection, mobile aquatic observation systems (Nittel, 2009)
  - Chemo-optical sensors: Real-time monitoring of air pollution (Caldararu et al., 2005)
- Remote sensing methods
  - Satellite imagery used to assess varying wavelengths or combinations of wavelengths in reflected light to detect changes in plant biomass and coverage
  - GIS systems used to integrate geospatial monitoring data for data management and analysis, visualization, and spatial and temporal modeling
- Nondestructive sampling and tracking
  - Hydro-acoustic sampling
  - Sampling of aquatic organisms position, density, and size
  - Radio and satellite tracking of fauna to assess habitat recolonization
  - Instruments for measuring vegetative water uptake

These technologies increase the ease of monitoring remote locations, make increasing sample sizes more feasible, and increase the detail of measurements. While cutting edge technologies are applicable in some situations, innovative technologies should not be chosen because they are cheaper or fancier. Technologies and equipment used for long-term and baseline monitoring need to be tested for reliability, accuracy, and robustness under field conditions. If emerging monitoring technologies do not meet these requirements, they should not be used for monitoring, regardless of cost savings or other benefits. Appropriate monitoring technologies need to be carefully chosen based on the present and future needs of the mining project.

REFERENCES

British Columbia Ministry of Environment, 2012. Water and Air Baseline Monitoring Guidance Document for Mine Proponents and Operators. Retrieved from: http://www2.gov.bc.ca/gov/DownloadAsset?assetId=E49A49E800814C8FB2D6868B7F119AD6.

Caldararu, F., Ionescu, C., Vasile, A., Caldararu, M., 2005. Chemo-optical sensor for toxic gases detection. In: Electronics Technology: Meeting the Challenges of Electronics Technology Progress, 2005. 28th International Spring Seminar on IEEE, pp. 71–75.

Carter, M.R., Gregorich, E.G., 2008. Soil Sampling and Methods of Analysis, second ed. CRC Press, Boca Raton, FL.

Department of Environment Climate Change and Water NSW, 2010. Environmental Compliance and Performance Report: Management of Dust from Coal Mines. Retrieved from: http://www.epa.nsw.gov.au/resources/licensing/10994coalminedust.pdf.

Environment Canada, 2011. An Integrated Oil Sands Environment Monitoring Plan. Retrieved from: http://www.ec.gc.ca/pollution/EACB8951-1ED0-4CBB-A6C9-84EE3467B211/Integrated%20Oil%20Sands_low_e.pdf.

Ge, L., Chang, H.-C., Rizos, C., 2007. Mine subsidence monitoring using multi-source satellite SAR images. Photogrammetric Engineering & Remote Sensing 73, 259–266.

Girard, J.M., 2001. Assessing and monitoring open pit mine highwalls. In: Proceedings of the 32nd Annual Institute of Mining Health, Safety and Research, Salt Lake City, UT.

Government of Alberta, 2012. Joint Canada/Alberta Implementation Plan for Oil Sands Monitoring. Retrieved from: http://www.ec.gc.ca/pollution/EACB8951-1ED0-4CBB-A6C9-84EE3467B211/Final%20OS%20Plan.pdf.

He, M., 2009. Real-time remote monitoring and forecasting system for geological disasters of landslides and its engineering application. Chinese J. Rock Mech. Eng. 6, 003.

He, M.C., Feng, J.L., Sun, X.M., 2008. Stability evaluation and optimal excavated design of rock slope at Antaibao open pit coal mine, China. Int. J. Rock Mech. Min. Sci. 45, 289–302.

Holmes, R.R., Terrio, P.J., Harris, M.A., Mills, P.C., 2001. Introduction to Field Methods for Hydrologic and Environmental Studies. U.S. Geological Survey. Retrieved from: http://pubs.usgs.gov/of/2001/0050/report.pdf.

Kissell, F.N., 2006. Handbook for Methane Control in Mining. National Institute for Occupational Safety and Health. Retrieved from: http://www.cdc.gov/niosh/mining/userfiles/works/pdfs/2006-127.pdf.

Lane, S.L., Flanagan, S., Wilde, F.D., 2003. National Field Manual for the Collection of Water-Quality Data. U.S. Geological Survey (USGS). Retrieved from: https://water.usgs.gov/owq/FieldManual/Chapter2/Chapter2_V2uncompressed.pdf.

Leading Practice Sustainable Development Program for the Mining Industry, 2009a. Airborne Contaminants, Noise and Vibration. Australian Department of Resources, Energy and Tourism. Retrieved from: http://www.industry.gov.au/resource/Documents/LPSDP/AirborneContaminantsNoiseVibrationHandbook_web.pdf.

Leading Practice Sustainable Development Program for the Mining Industry, 2009b. Evaluating Performance: Monitoring and Auditing. Australian Department of Resources, Energy and Tourism. Retrieved from: http://www.industry.gov.au/resource/Documents/LPSDP/EvaluatingPerformanceMonitoringAuditing_web.pdf.

Long, F., Zhu, A., Shi, H., 2013. Recent advances in optical biosensors for environmental monitoring and early warning. Sensors 13, 13928–13948.

Nittel, S., 2009. A survey of geosensor networks: advances in dynamic environmental monitoring. Sensors 9, 5664–5678.

Rantz, S.E., 1982. Measurement and Computation of Streamflow: Volume 1. Measurement of Stage and Discharge. U.S. Geological Survey. Retrieved from: http://pubs.usgs.gov/wsp/wsp2175/pdf/chapter10_vol2.pdf.

U.S. Environmental Protection Agency (U.S. EPA), 2001. Performing Quality Flow Measurements at Mine Sites. Retrieved from: http://nepis.epa.gov/Exe/ZyPDF.cgi/30002H0Y.PDF?Dockey=30002H0Y.PDF.

U.S. Environmental Protection Agency (U.S. EPA), 2002. EPA Air Pollution Control Cost Manual, sixth ed. Retrieved from: http://www.epa.gov/ttncatc1/dir1/c_allchs.pdf.

U.S. Environmental Protection Agency (U.S. EPA) Region 4, 2014. Operating Procedure: Soil Sampling. Retrieved from: http://www.epa.gov/region4/sesd/fbqstp/Soil-Sampling.pdf.

U.S. Environmental Protection Agency (U.S. EPA) Region 9, 1999. Field Sampling Guidance Document #1205 Soil Sampling. Retrieved from: http://www.epa.gov/region6/qa/qadevtools/mod5_sops/soil_sampling/r9soilsample_gui.pdf.

U.S. Geological Survey (USGS), 2014. How Streamflow is Measured. Part 2: The discharge Measurement. https://water.usgs.gov/edu/streamflow2.html.

Wagner, R.J., Mattraw, H.C., Ritz, G.F., Smith, B.A., 2000. Guidelines and Standard Procedures for Continuous Water-Quality Monitors: Site Selection, Field Operation, Calibration, Record Computation, and Reporting. US Department of the Interior, US Geological Survey.

Wyllie, D.C., Mah, C.W., 2004. Rock Slope Engineering: Civil and Mining, fourth ed. CRC Press.

CHAPTER 6

Environmental Auditing

6.1 INTRODUCTION

An important aspect of reducing the environmental impacts of mining and mineral processing operations is an environmental audit. In an environmental audit, environmental conditions are compared to established audit criteria or environmental goals. An audit is a critical stage in an environmental management system (EMS) as the "check" part of "plan-do-check-act" cycle used to minimize environmental impacts. Environmental risk and appropriate mitigation measures are established in an environmental audit and progress toward environmental goals are tracked. Audits should follow a systematic and documented process through which evidence is obtained and evaluated objectively (International Organization for Standardization (ISO), 2011; Leading Practice Sustainable Development Program for the Mining Industry, 2009; Ohio Aggregates & Industrial Minerals Association (OAIMA), 2014). In the United States, governmental organizations are audited according to the Generally Accepted Government Auditing Standards, known as the Yellow Book, issued by the Government Accountability Office. Internationally, organizations such as the International Organization of Supreme Audit Institutions and the Institute of Internal Auditors provide audit guidelines and pathways for certification. One of the most commonly accepted standards for auditing is ISO 19011, which includes environmental audits. This chapter will assist in the development of systematic procedures for environmental auditing.

6.2 TYPES OF ENVIRONMENTAL AUDITS

The frequency and type of environmental audit performed is dependent on whether the audit is voluntary, mandatory, or statutory (Table 6.1). The type of audit also affects the type of organization that can perform the audit. Mandatory or statutory audits require a independent auditor for credibility purposes. In addition, all auditors must have a level of competence and should be certified, if such certifications are available and common (Leading Practice Sustainable Development Program for the Mining Industry, 2009).

Environmental Impact of Mining and Mineral Processing
ISBN 978-0-12-804040-9
http://dx.doi.org/10.1016/B978-0-12-804040-9.00006-1

Table 6.1 Description of audit categories (Darnall et al., 2009; Leading Practice Sustainable Development Program for the Mining Industry, 2009)

Audit category	Description	Advantages	Disadvantages	Who performs audit?
Voluntary	An audit conducted without compulsion from authorities or required by law	Improves external image of company	Expensive and economically challenging if other companies do not participate; is not legally binding	Internal or independent organization
Mandatory	Required by license, permit, authority, legal powers	Gives assurance to stakeholders	Discrepancies from audit criteria could result in fines or closure	Independent organization
Statutory	Required by legislation	Gives assurance to stakeholders	Discrepancies from audit criteria could result in fines or closure	Independent organization

The main types of environmental audits are environmental management audits and environmental performance audits. While other audits are occasionally used, depending on the situation, management and environmental performance audits are widely used across all mining situations. These two audits can be used in conjunction with one another as environmental performance is closely intertwined with the environmental management system. An environmental audit can include many different subcategories of audits (Table 6.2).

6.3 PERFORMING AN ENVIRONMENTAL AUDIT

Throughout the auditing process, strict protocol must be followed. Standardizing the auditing process allows for easy comparison of performance between operations and allows for tracking of improvements. The major steps in the auditing process are as follows (International Organization for Standardization (ISO), 2011):

1. Initiate the audit
2. Prepare audit plan and activities

Table 6.2 Description of environmental audit subcategories

Audit type	Description
Environmental performance audit	• Verify the status of a mine with respect to predetermined environmental criteria. • This report should include the subject(s) to be audited, depth of audit, frequency and schedule of audit, and audit criteria.
Environmental management system audit	• Assesses if the mining company has implemented the EMS effectively throughout the mine. • Criteria for audit includes environmental policies, procedures, standards, and codes.
Compliance audit	• Assesses mine compliance against selected criteria derived from legislation, regulations, license, permits. • Can be a voluntary process for some mines, but is typically statutory.
Energy audit	• Assesses mine energy use based on tariffs investigations. • Includes major energy use equipment, operations, maintenance, and management processes.
Waste audit	• Assess all wastes generated. • Quantifies the type of waste and its composition, identifies reasons and factors for waste generation.
Environmental site audit	• Assesses contamination of a site for the purposes of real estate transactions, due diligence, or to meet regulatory requirements.
Environmental security audit	• Assesses vulnerability of hazardous material infrastructure, security management systems, and performs a gap analysis of environmental health and safety information.

Adapted from Leading Practice Sustainable Development Program for the Mining Industry (2009).

3. Conduct audit
4. Prepare report
5. Complete audit
6. Audit follow-up (if necessary)

The following audit steps are summarized from the International Organization for Standardization's ISO 19011:2011 Guidelines (2011).

Step 1: Initiate the Audit

An audit should be initiated by first contacting the auditee to establish a line of communication. Once authority to conduct an audit is received, information regarding the scope, objectives, methods, and documentation required for planning should be exchanged. Any accommodations necessary for site access, security, safety, guides, and observers should be arranged at this point. An audit may not always be feasible depending on the auditee cooperation, time and resource constraints, or availability of information for planning.

Step 2: Prepare Audit Plan and Activities

An audit plan should be developed by the audit leader to facilitate the conduct of the audit. An audit plan should include (International Organization for Standardization (ISO), 2011; Leading Practice Sustainable Development Program for the Mining Industry, 2009):

- Name and position of the auditee's representative
- Names of audit team members
- Audit objectives and scope
- Audit criteria
- Organizational and functional units to be audited
- Dates and places where the audit is to be conducted
- Expected time and duration for major audit activities
- Audit methodology
- Functions and/or individuals within the auditee's organization that have significant direct responsibilities regarding the audit
- Elements of the auditee's environmental and/or social management programs that are of high audit priority (based on risk)
- Procedures for auditing the auditee's management program elements, as appropriate for the auditee's organization
- Working and reporting languages of the audit
- Details of reference documents
- Schedule of meetings to be held with the auditee's management
- Report confidentiality requirements
- Report content, format and structure
- Expected date of issue and distribution of the report
- Document retention requirements

Any work documents, including checklists, sampling plans, and forms necessary for recording evidence should be prepared ahead of time to facilitate the auditing process. Once completed, audit plans should be reviewed and approved by the auditee before proceeding further. It is important that audit plans be flexible, as unforeseen changes may become necessary as the audit progresses and new information becomes available.

To adequately assess the success of an environmental management system and the environmental performance of a company, environmental indicators must be developed and certain areas must be emphasized in an audit. Indicators are simplifications of complex interactions between mining activities and environmental impacts. An indicator should increase the ease of quantification of performance and make the audit more informative to mining companies, regulatory agencies, and interested stakeholders. These indicators should be the focus of the audit. Sample questions pertaining to EMS audit parameters can be found at the end of this chapter.

Step 3: Conduct Audit

Meetings should be held at the start and end of the audit process. The opening meeting will help introduce all parties, ensure that the audit plan will be followed and activities carried out, establish lines of communication between the audit team and the auditee, and assign responsibilities for audit team members and any necessary guides.

Evidence must be collected for each of the indicators listed in Figure 6.1. Evidence can be rated based on the how the data was collected. The three primary methods of collecting evidence during an audit are:

- Review of documentation
- Observation of activities or situations
- Interview of appropriate personnel

Documentation can provide the highest standard of valid information, because the auditor does not have to rely on employee recollection, which

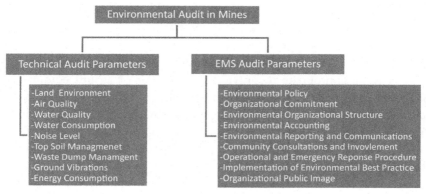

Figure 6.1 Break down of the main indicators used in an environmental audit (Kale and Yerpude, 2013).

may be incomplete or inaccurate. Documentation can include plans, records, procedures, policies, standards, drawings, contracts, specifications, permits, and licenses (International Organization for Standardization (ISO), 2011). If an effective EMS is in place, appropriate documentation should exist for all audit parameters identified during the previous step. When available, documentation is typically the first set of evidence collected during an audit and is checked to ensure it is complete, up-to-date, consistent, and correct (International Organization for Standardization (ISO), 2011). Reviewing documentation can be done remotely, or may take place on-site with auditee participation. However, documentation alone is no guarantee that the specified practices and policies are being following in the field.

Observations and interviews are conducted as follow-ups to ensure documented procedures and policies are being followed. On-site inspections and observations are necessary to determine compliance in areas without specific documentation or for cases when documentation does not accurately reflect operations. Observations should be thoroughly documented in detail and accompanied by a photographic record (if authorized) showing key areas of concern or non-compliance.

Interviewing personnel is especially important in determining the effectiveness of training and emergency response programs, operational compliance with documented procedures, communications between management and operators, and overall employee competence. Interviews should be performed in a manner that minimizes impacts to employees when possible (e.g., during normal working hours, close to their workplace).

During an audit, it may not always be practical to examine all available evidence due to factors such as time constraints, geographically dispersed operations, or the sheer volume of documentation available (International Organization for Standardization (ISO), 2011). In these cases, audit sampling may be appropriate to provide evidence for or against compliance. Sampling may introduce bias into the resulting audit conclusions if the selected sample is not representative. However, careful consideration of sampling methods, sample size, population composition, quality of the available data, and confidence level will help reduce bias and uncertainty in the results.

Following all evidence gathering activities, a closing meeting should be held among all parties. Closing meetings help auditors communicate the auditing findings and conclusions, as well as inform of any post–audit corrective actions or recommendations for improvement.

Step 4: Prepare Report

The audit report can include a large amount of detail on environmental processes or may only include deviations from audit criteria, depending on the purpose of the audit. These deviations can be ranked in order of environmental urgency. Recommendations for improvements can also be provided in the report. The contents of the audit report may include (Applied Environmental Management Consultants, 2012; International Organization for Standardization (ISO), 2011; Leading Practice Sustainable Development Program for the Mining Industry, 2009; SRK Consulting, 2009):

- Audit objectives and scope of work
- Audit team members and client
- Date and location of audit
- Audit methodology
  - Preparation and document review
  - Site visit activities
  - Reporting
- Site description and background information
  - Location
  - Geology
  - Operations, waste management, and water management
  - Controlling provisions, government requirements, and applicable leases or licenses
- Audit criteria
- Audit findings and evidence
  - Environmental aspects in compliance
  - Environmental aspects not in compliance
  - Observations and comments
  - Photographic documentation
- Conclusions and recommendations
- Statement of fulfillment of audit criteria
- Attachments
  - Terminology
  - Approval notification from controlling provisions

Steps 5 and 6: Complete Audit and Follow-up

After the audit report is issued, if all audit activities have been completed, then the audit is complete. Any records and documents should be stored or disposed of as previously agreed upon in the audit plan and any information obtained by the audit team during the audit should not be disclosed unless required by law.

Depending on the original audit objectives and findings, an audit follow-up may be appropriate. Any corrective actions or improvements agreed upon should be reported to the auditors and verified upon completion. Verification may require another audit.

6.4 STANDARDS FOR ENVIRONMENTAL AUDITING

Different standards for environmental auditing exist for different countries, making environmental auditing difficult within the current global economy. Global standards are preferred over country specific standards. The International Organization for Standardization (ISO) provides a global standard for environmental auditing, though not specifically for mining. These standards can be accessed for a fee through the ISO website. Some standards provide a certification that can be obtained voluntarily by a company or can be required by governments, while some standards provide a general guideline. The ISO standards below are the most up to date as of 2013 (International Organization for Standardization (ISO), 2013; Leading Practice Sustainable Development Program for the Mining Industry, 2009). New Zealand, Australia, and China have adopted these standards.

- ISO 14004:2004 provides guidance on the development and implementation of environmental management systems
  - ISO/AWI 14004: Environmental management systems – General guidelines on principles, systems and support techniques
- ISO 19011:2011: Guidelines for auditing management systems (International Organization for Standardization (ISO), 2011)

In addition to ISO guidelines, several other organizations promote standards that can be used when performing an audit (Leading Practice Sustainable Development Program for the Mining Industry, 2009).

- ASTM International, Standard E1528–06 Standard practice for environmental site assessments: transaction screen process, ASTM.
  - E1527–05 Standard practice for environmental site assessments: Phase I environmental site assessment process, ASTM.
  - E1903–11 (2011) Standard practice for environmental site assessments: Phase 2 environmental site assessment process, ASTM.
- Auditing and Assurance Standards Board 2008, ASAE 3100 Compliance engagements, AUASB, September 2008.
  - 2007, ASAE 3000 Assurance engagements other than audits or reviews of historical financial information 2007, AUASB, July 2007.
- Environment Protection Authority South Australia 2003, Protection for voluntary environmental audits, SA EPA 013/03, September 2003.
- Environment Protection Authority Victoria 2007, Environmental auditor guidelines for conducting environmental audits, Publication 953.2, August 2007.
- World Bank 1995, "Environmental Auditing," Environmental Assessment Sourcebook Update 11, Environment Department, Washington DC.
- World Bank Group 1998, Environmental audits in industrial projects, Pollution Prevention and Abatement Handbook, July 1998.

6.5 AUDITING SYSTEM CHECKLISTS

When performing an audit on an environmental management system, reviewing documents may not be enough to evaluate the effectiveness of the management system. Aspects such as communication, training, and community involvement need to be evaluated. A list of questions (Kale and Yerpude, 2013) has been developed to assist in the data collection aspect of the environmental management system audit (Table 6.3).

When performing an environmental audit, a checklist can be prepared ahead of time and can make the auditing process more efficient. Table 6.4 is an example developed by the Ohio Aggregates and Industrial Minerals Association (2014).

Table 6.3 An example of a checklist used for environmental management system audits (Kale and Yerpude, 2013)

Environmental policy	Follow-up needed:
Does the organization have a comprehensive environmental policy?	
Does the policy have sufficient scope to include all of the organization's management units?	
Does the organization have a separate environmental policy for each mining unit?	
Does the policy address and is it appropriate for the environmental impacts due to mining activities, considering the nature of operations?	
Does the policy establish a commitment to continual improvement and prevention of pollution?	
Does the policy establish a commitment to comply with relevant environmental legislation and regulations and with other requirements?	
Does the policy provide the framework for setting and reviewing environmental objectives and targets?	
Does the policy provide directions for documenting, implementing, and maintaining the environmental policy?	
Are mining employees aware of the environmental policy of the organization?	
Does the policy have provisions for making policies available to the public?	

Organizational commitment	Follow-up needed:
Has the principal environmental policy been approved at the highest level of management?	
Does management sufficiently allocate and spend funds on environmental management of the mining activities?	
Has a responsible management representative been assigned to ensure implementation of EMS and reporting to the top management?	
Does the organization have a procedure to identify the environmental aspects of mining activities and does it consider the impacts of those aspects in setting environmental objectives?	
Has the organization developed a program for achieving its environmental objectives and targets?	
Has responsibility been assigned to each level to achieve objectives and targets?	
Are there internal meetings to evaluate the environmental performance of the mining unit?	
Has the organization given topmost priority to comply with regulatory standards with respect of the environment?	
Does the organization assign personnel to various training courses to update their technical skills?	
Does the organization have R&D programs for abatement of mining-related environmental pollution?	

	Follow-up needed:
Environmental organization and structure	
Are there separate environment management divisions dedicated for each mining unit?	
Is the head of the environment management division part of the decision-making process for the mining unit?	
Are the employees in the environmental management division technically and professionally competent?	
Are the roles, responsibilities, and authorities clearly defined for the persons appointed in the environmental management divisions?	
Are adequate human, financial, and technological resources provided to this division?	
Are the personnel appointed in the division aware of scientific procedures for carrying out environmental investigations?	
Are the personnel appointed in the division aware of relevant environmental legislation and terms and conditions of the various environmental related approval/clearances received by the organization?	
Are the environment related statutory and non-statutory registers, documents, returns, etc., maintained and updated?	
Does an environmental manual contain a compilation of all EMS components, procedures, regulations, legal and technical documents, etc., and is it maintained and kept up to date?	
Do personnel posted in the environmental management division periodically undergo environment specific training programs, refresher courses, workshops, seminars, symposia, etc.?	

	Follow-up needed:
Environmental accounting	
Is a separate budget allocated for the environmental management of mining operations?	
Is the allocated budget for environmental management being utilized to the extent of 90% or more?	
Is the cost incurred for environmental management of mining operations separately accounted toward the environmental head?	
Does the percentage of cost incurred toward environment management to the total cost expenditure of the unit exceed 5%?	

Continued

Table 6.3 An example of a checklist used for environmental management system audits (Kale and Yerpude, 2013)—cont'd

Training	Follow-up needed:
Does the mining organization have a full-fledged "training center" for training of employees?	
Have responsibilities been assigned for reviewing, updating, and overseeing implementation of training procedures?	
Is the mining company providing environmental training for employees?	
Are outside experts invited to impart training on environmental aspects?	
Is there a system for identification and evaluation of training needs?	
Have line supervisors and managers undergone periodic work specific environment related training?	
Are online/practical aspects of the work stressed in the training programs?	
Have supervisors/managers of the concerned division undergone training on the environmental legislation applicable to mining industry?	
Have the employees involved in monitoring of environmental performance undergone training in that particular field?	
Is the mine management organizing programs to observe various environmental related functions such as World Environment Day, Earth Day, etc.?	

Environmental reporting and communications	Follow-up needed:
Does the mining unit have procedures for internal communication on environmental matters between the various levels and functions of the organization?	
Has the mining organization complied with the mandatory disclosure of environmental activities?	
Has the mining company voluntarily disclosed information about their environmental activities in addition to the mandatory disclosure?	
Are environmental reports communicated to the public at least once a year?	
Has the mining organization allowed for receiving, documenting, and responding to communications on environmental matters from external parties?	
Has responsibility been assigned for reviewing, updating, and overseeing the implementation of communications procedures?	

	Follow-up needed:
Are environmental activities featured in reports and news issued by the mining company?	
Are environmental reports presented by the mining company in various functions organized by the company?	
Does the mining company highlight their environmental activities in the form of technical papers in seminars/symposia?	
Has the mining company allowed NGOs to visit their mining operations and environment related programs?	
Community consultations and involvement	
Has the mining company voluntarily disclosed information about their environmental activities in addition to the mandatory disclosure?	
Is mine management aware of the importance of community consultation for the success of the project?	
Have community consultations begun in initial stages of the mine development?	
Does mine management have an appointed "liaison officer" for the purpose of community relations?	
Is the liaison officer aware of his roles and responsibility for community relations?	
Has the liaison officer identified existing and forecasted potential concerns of the community for mine operations?	
Are all community concerns, complaints, and inquiries being addressed promptly?	
Does the liaison officer maintain a register of public concerns, complaints, and inquiries?	
Are community consultations and involvement still a frequent, ongoing activity?	
Is senior management aware of public concerns and the local mine management's response to those concerns by way of checking registers, meetings, site visits, etc.?	

Continued

Table 6.3 An example of a checklist used for environmental management system audits (Kale and Yerpude, 2013)—cont'd

	Follow-up needed:
Operational and emergency response procedures	

Has mine management identified all mining activities that have an environmental impact?

Are standard operational procedures consistent with best environmental practices that have been established?

Have environmentally sound and standard operational procedures been documented?

Are standard operational procedures being reviewed, amended, and supplemented periodically?

Are the staff and officers involved in day-to-day working acquainted with the standard operational procedures?

Have potential risks and accidents been identified?

Have emergency response procedures been established to deal with any unforeseen events?

Are periodic tests of procedures, such as drills, exercises, or mock operations being conducted?

Has mine management assigned responsibility to administer and update the operational and emergency response procedures?

Have environment related scientific studies or risk assessment studies been performed?

	Follow-up needed:
Implementation of environmental best practices	

Has the mine management adopted innovative methods for better environmental management?

Is the mining company involved in carbon trading?

Has the mining unit adopted international standards, such as ISO standards, for better environmental management?

Does the mine management carry out R&D programs for environmental problems related to mining via either in-house research or collaboration with scientific institutions?

Has mine management outsourced technical consultancy to mitigate environmental problems?

Does the organization have a system to assess the appropriateness of any particular procedure, system, or technology in order to evaluate it as environmentally best practice?

	Follow-up needed:
Does the mining company have any schemes to encourage employees to adopt innovative methods?	
Does the mining company share and circulate knowledge of the advantages of adoption of new innovative methods to other mining companies?	
Has the mining company received recognitions at the national or international level for its adoption of environmental best practices?	
Has the mining company received patents for developing environmental best practices?	

Organizational public image

	Follow-up needed:
Has mine management adopted international standards such as ISO?	
Has the mining unit provided adequate compensation to the project-affected people and communities?	
Did the mining project come as a boom regarding employment for the local population?	
Has the reporting period for an audit been free from any environment related complaints from local community members?	
Has the company received any awards during the reporting period?	
Has the financial status of the company shown positive growth during the reporting period?	
Have the shares of the mining company been positive during the reporting period?	
Has mine management promoted community development in or around the mining site?	
Does the mining company highlight its good work in the economic, social, environmental, and technological fields through various means?	
Does the attrition rate remain less than 12% during the reporting period?	

Table 6.4 Example checklist used for mining audits. This list is specific to mining in the United States and would need to be adapted to another country's permits, regulations, and requirements (Ohio Aggregates & Industrial Minerals Association (OAIMA), 2014)

Air quality	Follow up needed:
Has an air quality permit been issued for this facility? Permit number(s):	
Date of issuance: Expiration:	
Are emissions being controlled on-site?	
Are entrance roads clean of debris?	
Are trucks covered and free of loose stone and sand?	
Are notices posted to cover loads and check tailgates?	
Are permit restrictions on production being met? How is this documented?	
Are permit restrictions on sales being met? How is this documented?	
Are permit conditions for water application rates for roadways being met? How is this documented?	
Are permit conditions for the plant dust control being met, i.e., water sprays, bag houses, opacity limits? How is this documented?	
Do operators know opacity limits?	
Number of people at this facility certified to do opacity readings:	
Are permit restrictions for generators being met, i.e., amount of fuel consumed, opacity limits, etc.? How is this documented?	
Are inspections and maintenance of air pollution control equipment scheduled on a regular basis? How is this documented?	
Are all required records adequate, accurate, and readily available for inspection?	
What contingency plan in place for malfunction of emission control equipment?	
Have all additions or modifications of this plant been permitted?	
Are there pending modifications?	
Are submittals to regulatory agencies done on a timely basis?	

	Follow-up needed:
Water quality	
Has a wastewater discharge permit (general or individual) been issued to this facility? Permit number(s):	
Date of issuance: expiration:	
Is process water being controlled and contained in sediment ponds?	
Does the final effluent from the sediment ponds appear to be free from floating solids, visible foam (only trace amounts allowed), and oil? How is this documented?	
Are samples being collected, handled, and analyzed properly? How is this documented?	
Are discharge monitoring reports on-site?	
Does the effluent meet the permit limits for turbidity, pH, TSS, etc.?	
Are all required records adequate, accurate, and readily available for inspection?	
What contingency plans are in place for malfunction of wastewater control equipment?	
Have all additions or modifications of this plant been permitted?	
Are there pending modifications?	
Are submittals to regulating agencies done on a timely basis?	
Is storm water regulated under this permit?	

	Follow-up needed:
Storm water	
Has a storm water discharge permit been issued to this facility? Permit number(s):	
Date of issuance: expiration:	
Does the facility have a storm water management (pollution prevention) plan?	
Date of plan: does the plan need to be amended? By what date?	
Are best management practices being employed to control storm water? How is this documented?	
Are samples being collected, handled, and analyzed properly? How is this documented?	
Are storm water discharge monitoring records on-site?	
Does the effluent meet the permit limits for turbidity, pH, TSS, etc.?	

Continued

Table 6.4 Example checklist used for mining audits. This list is specific to mining in the United States and would need to be adapted to another country's permits, regulations, and requirements (Ohio Aggregates & Industrial Minerals Association (OAIMA), 2014)—cont'd

	Follow-up needed:

Are all required records adequate, accurate, and readily available for inspection?

What contingency plans are in place for the malfunction of storm water control equipment?

Are submittals to regulatory agencies done on a timely basis?

List where silt fences are needed:

List where rip rap is needed:

List where ground cover is needed:

List where sediment pond maintenance is needed:

Water resources

Non-potable: has a water withdrawal permit (surface or groundwater) been issued to this facility?

Permit number(s):

Date of issuance: expiration:

Does the facility have a water conservation plan?

Date of plan: does the plan need to be amended? By what date?

Are wells monitored? Are pumping volume records on-site? How is this documented?

Are all required records adequate, accurate, and readily available for inspection?

Are submittals to regulatory agencies done on a timely basis?

Potable water: has a drinking water permit been issued to this facility? Permit numbers:

Registered well(s):

Date of issuance: expiration:

Does the facility have a drinking water testing program?

Are samples being collected, handled, and analyzed properly? How is this documented?

Does the drinking water meet the limits for all required testing parameters? How is this documented?

Are inspections and maintenance of treatment systems and requirements scheduled on a regular basis? How is this documented?

Are submittals to regulatory agencies made on a timely basis?

What contingency plan is in place for the malfunction of drinking water control equipment?

Is bottled water provided at this facility?

	Follow-up needed:
Surface mining permit	
Has a mining permit been issued to this facility? Permit number(s):	
Number of permitted acres: number of disturbed acres:	
Is the permitted area flagged or marked off?	
Does the facility have a mining land use plan? Date of plan:	
Is the facility complying with the plan? How is this documented?	
Are ID numbers and contact names in an office or on signs?	
Are any changes planned?	
Are submittals to regulatory agencies done on a timely basis?	

	Follow-up needed:
Site of archeological significance	
Has a historical or cultural significance permit been issued for this facility? Permit number(s):	
Effective date: expiration date:	
Are regulated wetlands or "waters of the state" on this site?	
Are these in current or future mining areas?	
Are best management practices being employed to protect stream buffers and maintain barge facilities?	
How is this documented?	
Have endangered or threatened species been identified on this site?	
Are these species currently being protected from harm?	
Have archeological sites been identified in current or future mining areas?	
Are these sites currently protected from disturbance?	

Continued

Table 6.4 Example checklist used for mining audits. This list is specific to mining in the United States and would need to be adapted to another country's permits, regulations, and requirements (Ohio Aggregates & Industrial Minerals Association (OAIMA), 2014)—cont'd

Aboveground and underground storage tanks	Follow-up needed:
Has an aboveground storage tank been permitted or registered at this facility? Permit or registration number(s):	
Date of issuance: expiration:	
Does the facility have a stamped spill prevention, control, and countermeasure plan for petroleum products?	
Date of plan: does plan need to be amended? By what date?	
Does the facility have a spill plan for chemical storage?	
Date of plan: does plan need to be amended? By what date?	
Is secondary containment provided for all aboveground tanks?	
Are inventory records maintained?	
Is Phase II gasoline vapor recovery provided where required?	
Are tanks equipped with anti-siphon and overfill protection?	
Are inspections and maintenance of storage systems scheduled on a regular basis? How is this documented?	
Is inspection follow up adequate?	
Is effluent from fuel containment dikes or treatment systems free of oil sheen and periodically tested to verify TPH, etc. levels? How is this documented?	
Is effluent from other containment dikes or treatment systems free of product? How is this documented?	
Are drain valves maintained in locked position except during monitored release?	
Are spills cleaned up when they occur?	
Are adequate spill supplies available in an accessible location?	
Are all required records adequate, accurate, and readily available for inspection?	
Are submittals to regulatory agencies done on a timely basis?	
What contingency plans are in place for the malfunction of control equipment?	
What type of security systems is in place to prevent outside persons from tampering with aboveground tanks?	

Continued

What measures are taken to prevent vandalism to tanks?

Has an underground storage tank been permitted or registered at this facility? Permit or registration number(s):

Date of issuance: expiration:

Are the tanks and piping equipped with corrosion prevention, release detection, and overfill protection? How is this documented?

Are tests done as required if the tanks or piping are catholically protected? How is this documented?

Are underground storage tank inventory records routinely reconciled (stick vs meter)? How is this documented?

Are tank tightness tests performed as required? How is this documented?

For tanks that were closed, have samples been collected, handled, and analyzed properly? How is this documented?

Are all required records adequate, accurate, and readily available for inspection?

What contingency plans are in place for the malfunction of control equipment?

Is the facility participating in a trust fund to assist with leaking tank cleanup costs?

Are submittals to regulatory agencies done on a timely basis?

Is used oil in properly labeled containers?

Is used oil being properly manifested?

What is the registration ID number of the used oil hauler?

Is a solvent rinse basin available for parts cleaning at this facility?

What is the procedure for the disposal of solvent at this facility?

What is the name of this solvent?

What is the approximate maximum number of pounds on site at any time?

Is waste solvent being properly manifested?

Name and telephone number of waste solvent hauler:

Are there other potential hazardous wastes?

Are hazardous waste containers being properly dated and labeled?

Are hazardous wastes properly manifested?

Table 6.4 Example checklist used for mining audits. This list is specific to mining in the United States and would need to be adapted to another country's permits, regulations, and requirements (Ohio Aggregates & Industrial Minerals Association (OAIMA), 2014)—cont'd

Are hazardous wastes volumes sufficient for the facility to be determined a small or large quantity generator? How is this documented?

Is this facility served by municipal sewerage treatment?

Are there pretreatment standards?

Are these standards being met? How is this documented?

Does the facility have an approved septic system?

Does the facility have a holding tank for gray water or sewage?

Are holding tanks emptied regularly?

Is there a contract for solid waste pickup?

Is final disposal in a suitable landfill?

Name and telephone number of solid waste hauler:

What is the registration ID number of the used oil hauler?

What is contained in the spare materials yard(s)? Transmissions? Rear ends? Engine oil or antifreeze drums? Transfer cases? Air conditioners? Transformers?

Are there polychlorinated biphenyl (PCB) transformers at this facility?

Are inspections and maintenance of the transformers scheduled on a regular basis to prevent leakage? How is this documented?

Are all required records adequate, accurate, and readily available for inspection?

Are submittals to regulatory agencies done on a timely basis?

Is there a program to remove these transformers with a non-PCB type? Describe:

Are there asbestos containing materials at this facility?

Are inspections and maintenance of these materials scheduled on a regular basis to prevent particles from becoming airborne? How is this documented?

Are all required records adequate, accurate, and readily available for inspection?

Are submittals to regulatory agencies done on a timely basis?

Is there a program to remove these materials? Describe:

Are there lead containing materials at this facility?

Are adequate precautions taken to prevent environmental contamination? How is this documented?

Is there a program to remove these materials? Describe:

Are chlorofluorocarbons (CFCs) or other refrigerants being properly collected and disposed?

Chemical handling

Are material safety data sheets (MSDSs) available for all chemicals for industrial use?
Are all containers labeled?
Are pesticides, herbicides, or rodenticides used at this facility?
Are adequate precautions taken to prevent unwanted environmental consequences?

Follow-up needed:

Noise

Are there restrictions on noise levels at this facility?
Does the facility meet these levels? How is this documented?

Follow-up needed:

Contractors

Are contract drillers, blasters, mechanics, strippers, etc. following safety and environmental regulations and company policies?
Are environmental clauses written into contracts?
Are company policies and guidelines made available to contractors?
Are company regulatory requirements made available to contractors (e.g., mining land use plans to stripping contractors)?
Do contract drillers maintain good dust control?
Do contractors handle wastes properly?
Do contract strippers follow best management practices for sediment and erosion control?
Are there current work clearance or hold harmless certificates on file for these crews?

Follow-up needed:

Continued

Table 6.4 Example checklist used for mining audits. This list is specific to mining in the United States and would need to be adapted to another country's permits, regulations, and requirements (Ohio Aggregates & Industrial Minerals Association (OAIMA), 2014)—cont'd

Community	Follow-up needed:
Is the facility adequately screened from the community?	
Is the facility actively involved in the community? Describe programs:	
Are all blasts monitored? How is this documented?	
Do blast records reveal any problems?	
Are close proximity blasting guidelines being followed at this facility?	
Are historic structures in the immediate vicinity?	
Does this require modifications to the operations (e.g., reduced blasting limits)?	
Has the facility received any complaints during the previous 3 years? List:	
Describe the procedures for addressing an environmental complaint:	
Have local fire department and emergency response teams toured the facility?	
Are community right-to-know reports and MSDSs submitted to state and local authorities?	

Zoning	Follow-up needed:
Are local zoning or conditional use restrictions in place for this facility?	
Is the facility complying with these restrictions? How is this documented?	
Are required reports submitted on a timely basis?	

Training	Follow-up needed:
Do employees know the environmental policy and how it pertains to their jobs?	
How is the environmental policy communicated?	

Are employees aware of the environmental impacts of the operations and their role in minimizing these impacts?

Do the employees know the consequences of not following company policies?

Are employees trained on site-specific spill prevention, control, and counter measure plans?

Are employees trained on how to clean up spills of fuels and other hazardous materials?

Is training provided on the safe handling of hazardous materials used at workstations?

Are employees informed where MSDSs are kept?

Are instructions given to employees on when and how to use personal protective equipment (PPE)?

Are instructions given to employees on who to notify in the event of an emergency?

Has management received crisis management training?

Are training records maintained? How is this documented?

Follow-up needed:

Documentation

Are forms for regulatory or other compulsory reporting controlled to ensure correct version is used?

Are record maintenance policies documented and followed?

Are records used for regulatory or other compulsory reporting protected from damage or loss?

Are current copies of environmental permits maintained at the facility?

Are pertinent environmental regulations readily available?

Follow-up needed:

Enforcement actions

Have there been any Notice of Violation or other enforcement actions during the previous 3 years? Describe:

Are there any open notices of violation?

Is the facility meeting their compliance schedule? How is this documented?

Follow-up needed:

Monitoring and measurement

Are the seismographs used to monitor blasts calibrated? How is this documented?

Are the pH meters used for regulator reporting calibrated? How is this documented?

Follow-up needed:

Continued

Table 6.4 Example checklist used for mining audits. This list is specific to mining in the United States and would need to be adapted to another country's permits, regulations, and requirements (Ohio Aggregates & Industrial Minerals Association (OAIMA), 2014)—cont'd

	Follow-up needed:
Are analytical balances used for regulatory testing calibrated? How is this documented?	
Are personnel performing New Source Performance Standard (NSPS) Subpart OOO testing certified? How is this documented?	
Are other equipment used for regulatory or compulsory monitoring calibrated? How is this documented?	
Environmental performance indicators	
Have site-specific environmental goals been set? Describe:	
Are environmental programs in place to assist in meeting these goals?	
Are there any resources that are needed to accomplish these goals?	
Are corporate key environmental performance indicators being tracked at this facility?	
Are other pollution prevention or environmental enhancement actives occurring at this facility? Describe:	
Are there other resources or support needed by the facility to ensure environmental compliance and ongoing improvements? List:	

REFERENCES

Applied Environmental Management Consultants, 2012. Mt Arthur Coal Independent Environmental Audit – November 2012. Retrieved from: http://www.bhpbilliton. com/home/society/regulatory/Documents/Mt%20Arthur%20Coal%202012/Mt%20 Arthur%20Coal%20Independent%20Environmental%20Audit%202012.pdf.pdf.

Darnall, N., Seol, I., Sarkis, J., 2009. Perceived stakeholder influences and organizations' use of environmental audits. Account. Org. Soc. 34, 170–187.

International Organization for Standardization (ISO), 2011. ISO 19011 Guidelines for Auditing Management Systems, second ed. Retrieved from: http://www.cnis.gov.cn/ wzgg/201202/P020120229378899282521.pdf.

International Organization for Standardization (ISO), 2013. ISO-1400 Environmental Management. Retrieved from: http://www.iso.org/iso/home/standards/management-standards/iso14000.html.

Kale, Y.G., Yerpude, R.R., 2013. Environmental audit system for opencast mines-monitoring approach for green mining. J. Min. Metals. Fuel. 61, 5–18.

Leading Practice Sustainable Development Program for the Mining Industry, 2009. Evaluating performance: monitoring and auditing. Australian Department of Resources, Energy and Tourism. Retrieved from: http://www.industry.gov.au/resource/ Documents/LPSDP/EvaluatingPerformanceMonitoringAuditing_web.pdf.

Ohio Aggregates & Industrial Minerals Association (OAIMA), 2014. Environmental Compliance Manual. Retrieved from: http://oaima.org/aws/OAIMA/pt/sp/manual.

SRK Consulting, 2009. Environmental Audit of the Greens Creek Mine – Final Report. Retrieved from: http://dnr.alaska.gov/mlw/mining/largemine/greenscreek/pdf/gcaudit 2009.pdf.

CHAPTER 7

Mitigation Measures and Control Technology for Environmental and Human Impacts

7.1 INTRODUCTION

Mitigation measures and control technology are important aspects of environmental management systems and are vital for the minimization of negative environmental and human health impacts from mining and mineral processing. Mitigation of environmental impacts in one area can have widespread benefits on several other environmental attributes and thus, mitigation should always be considered, regardless if the potential impacts may seem insignificant. Mitigation of negative environmental impacts should adhere to the following broad categories of environmental commitment: avoidance, minimization, restoration, repair, reducing impact over time, and compensation (Bellamy and Nicholson, 2012). Mitigation measures should always strive for the avoidance of negative impacts, followed by minimization if avoidance is not feasible, and so forth. In this way, the highest level of mitigation is assured. Mitigation and control measures should be incorporated when evaluating alternatives and assessing environmental impacts and controls schemes. By implementing mitigation measures and control technology early in a mine's life, costly cleanup and remediation operations can be avoided.

7.2 AIR POLLUTION MITIGATION AND CONTROL

Air pollution from mining and mineral processing operations can affect local and regional air quality. The majority of air pollution concerns consist of methane, particle emissions (dust), and other various gaseous emissions. It is important to control and mitigate air pollution at the source, as there is no feasible method to treat and remove pollutants once they become dispersed in the atmosphere.

Environmental Impact of Mining and Mineral Processing
ISBN 978-0-12-804040-9
http://dx.doi.org/10.1016/B978-0-12-804040-9.00007-3

Mitigation of Particulate Matter

Throughout the life cycle of a mine, mitigation techniques should be used to reduce PM creation and transport. In the United States, mining companies are required to control and minimize fugitive dust in order to meet state and national air quality standards. Many states require the development of fugitive dust control plans to assist in dust mitigation.

A list of problem areas concerning particulate emissions and mitigation techniques is provided below to aid in choosing mitigation measures throughout the life of the mine (Leading Practice Sustainable Development Program for the Mining Industry, 2009).

- Mine planning
  - Consider the distance from buildings to mining activities as well as prevailing winds (distance and whether mining site is up or down wind from development)
  - Size of the site: The larger the site, the farther pollutants will have to travel to leave the site, decreasing dust impacts on neighboring communities
  - Consider local hills or rises: Avoid placing a mining site on a hill. Placing the site on downwind side of a hill means that the wind will not be as strong
  - Consider existing flora: Trees create windbreaks and can collect dust
  - Road design: Consider the location and quality of design
  - Plan emergency tactics that can be implemented in unfavorable emission conditions
    - Increased monitoring
    - Watering
    - Limiting or ceasing activities
- Mine operation
  - Moisture content of soil needs to be above the amount that produces dust but below the level where it hinders vehicle and personnel movement
  - Lower PM levels improve safety through increasing visibility
- Transfer and materials handling plants
  - Verify material has a relatively high moisture content, including prior to being separated into course and fine material
  - Make sure all operating equipment is enclosed and sealed if possible
  - Reduce air flow in and around conveyance areas and stockpiles to prevent small particles from being carried into air
  - Where flow direction of raw material changes, use impact plates and air stream flow to pick up fine particles

- Add moisture or appropriate chemical binders to the raw material stream before dumping, tipping, conveying, or transferring
- Conveyors
  - Install wind barrier over conveyor structure
  - Modify conveyor belt to decrease vibrations
  - Clean conveyor belt
- Transfer points
  - Use a hood and spoon chute rather than typical box chute
  - Cascade rock box chute: Should not be used for fine material, decrease energy of impact by decreasing drop height
- Moisture splitting
  - After coarse and fine particles are separated, add moisture to coarse particles
- Roads
  - Selection of appropriate road surfaces
  - Use of water treatment or a chemical treatment such as calcium chloride
  - Control vehicle speed
  - Spray idle open areas with seeding mulch to build up vegetation
  - Daily roadway sweeping
  - Immediately cleanup of any spills
- Stockpiles and stockyards
  - Stackers must have misting sprayers installed onto the boom, directed to form a curtain around falling materials to trap fine particles
  - Stockpiling should be performed to minimize drop distances
  - Mobile reclaimers need two sets of nozzles: One at the face of stockpile and one behind the cutting wheel
  - Drum reclaimers need sprays although no water should be directed toward equipment
  - Stockpiles should be sprayed before large wind events
  - Chemical encrusting agents can be sprayed onto stockpiles using water cannons or water trucks

Mineral and metal processing can also contribute to particulate emissions. Typical areas of concern and mitigation options are provided in the following list:

- Roasting
  - Gases from roasting should be filtered
- Sintering
  - Strong gases from the front end of the sintering machine should be filtered

- Weak gases from the back end of the sintering machine should be filtered
- The building, sintering machine, and sinter cake crusher should be checked for fugitive leaks
- Blast furnace
  - Gases from the blast furnace should be filtered
  - Blast furnace, furnace covers, and charge cars should be checked for fugitive leaks
  - Fuel combustion products can contribute to particulate emissions
  - Periodic cleanout should be done carefully to minimize particulate emissions
  - Waste slag processing and piles may contribute to fugitive dust

Particulate Matter Control Technology

Engineering controls for dust management can be generally divided into three categories: containment, suppression, and collection (Department of Environment Climate Change and Water NSW, 2010). Containment systems are mechanical control methods used to prevent dust dispersion and are the first steps taken to minimize dust pollution (Department of Environment Climate Change and Water NSW, 2010). Containment can be relatively simple, such as enclosing conveyors, minimizing dump heights and rates from trucks, and resurfacing of high-traffic areas. Containment may also involve sealing or planting vegetation on stockpiles, waste piles, and tailings to stabilize the soil, installing wind breaks (e.g., trees, shrubs, artificial barriers), or building enclosures around large dust generating equipment like screens and crushers.

Suppression systems are designed to reduce fugitive dust and PM through the use of water or chemical application (Department of Environment Climate Change and Water NSW, 2010). The primary method of suppression is to apply water at all stages of construction, operation, and rehabilitation as required. Water can be applied using a variety of methods, including water cannons, water trucks, sprinkler systems, or automated sprinkler systems that activate depending on wind speed and vibrations. Chemicals additives such as surfactants or foaming agents may be used to increase wetting efficiency (Martin Marietta Laboratories, 1987). During rehabilitation, most particulate emissions come from wind erosion, so rapid revegetation is essential to reduce PM emissions (Leading Practice Sustainable Development Program for the Mining Industry, 2009).

Particulate matter collection systems are designed to remove PM suspended in the air and typically involve an exhaust hood to collect emissions, ductwork to transport the emissions, a PM collector, a fan to provide necessary airflow, and the discharge of clean air back into the environment (Department of Environment Climate Change and Water NSW, 2010; Martin Marietta Laboratories, 1987). Various types of PM collection systems are presented in Table 7.1. Fabric filters, or baghouses, operate by forcing PM filled air through fabric bags, collecting PM on the upstream side of the bags and releasing the filtered air to atmosphere. Fabric filters generally have removal efficiencies of >99% and can remove particles smaller than 1 micron due to the dust cake layer that forms on the fabric (U.S. Environmental Protection Agency (U.S. EPA), 2002). Fabric filters are generally designed to be cleaned through shaking, reverse air or pulse-jet cleaning, or sonic cleaning; fabric material should be selected to prevent particle adherence while maintaining performance (U.S. Environmental Protection Agency (U.S. EPA), 2002).

Wet scrubbers use liquid droplets to remove PM via impaction, interception, absorption, and diffusion. Removal efficiencies decrease with PM size, but generally range from 99% to 40% depending on the system. Types of wet scrubbers include spray towers, cyclonic spray towers, dynamic scrubbers (or mechanically-aided scrubbers), tray towers, Venturi scrubbers, orifice scrubbers (or impaction scrubbers), packed tower scrubbers, condensation scrubbers, and charged scrubbers (U.S. Environmental Protection Agency (U.S. EPA), 2002). Wet scrubbers are used for stationary point sources and while particularly effective at removing sticky, hygroscopic, combustible, and corrosive materials, are limited to low flow emissions (U.S. Environmental Protection Agency (U.S. EPA), 2002).

Electrostatic precipitators utilize electrostatic forces to collect particulate emissions onto collector plates. Several variants include the plate-wire precipitator, flat plate precipitator, tubular precipitator, wet precipitator, and the two-stage precipitator. In the flat plate precipitator, the most common type, emissions flow through metal plates and electrodes where particles and ions can collect on the walls (U.S. Environmental Protection Agency (U.S. EPA), 2002).

An automated particulate emissions control system can be integrated into the mining process to aid in the application of mitigation techniques. This automated control system can incorporate data from emissions monitoring, weather prediction, local current weather, material characteristics, and mining schedules for raw material, waste, and soil movement

Table 7.1 Types of particulate matter collector systems (Martin Marietta Laboratories, 1987; U.S. Environmental Protection Agency (U.S. EPA), 2002)

Type of PM collector	Subtypes	Description
Inertial separators	Settling chambers	Large ductwork box designed to reduce air speed for gravity separation, primarily used as a precleaning step
	Baffle chambers	Chamber with a fixed baffle plate causing a change in flow direction, primarily used as a precleaning step
	Centrifugal collectors	Cone shaped chamber causes cyclonic action to separate dust in an outer vortex, with fresh air in the inner vortex
Fabric collectors		Fabric bags act as filters. Filters can be made of cotton, synthetic, or glass fibers, filters dust via inertial collection, interception, Brownian movement, and electrostatic forces, very efficient and cost effective
Wet scrubbers		Scrubbing liquid (typically water) is brought into contact with the contaminated air, removes dust via inertial impaction, interception, diffusion, and condensation nucleation
Electrostatic precipitators		Uses electrostatic forces to separate dust from air, particles are ionized and adhere to electrodes as contaminated air is passed through
Unit collectors	Fabric collectors Cyclone collectors	Contain dust at its source, small, self-contained, portable units for operations such as conveying, screening, and crushing

to guide the set up and design of automated water application, sprinkler systems, chemical treatment systems, belt wash systems, and particulate emissions removal systems (Leading Practice Sustainable Development Program for the Mining Industry, 2009).

Mitigation of Volatile Organic Compound Emissions

Effective mitigation of VOC emissions during filling and storage relies on the implementation of proper maintenance and handling procedures and vapor recovery equipment. Proper underground storage tank filling methods, such as submerged or balanced submerged filling, can significantly reduce emissions when compared to splash filling (U.S. Environmental

Protection Agency (U.S. EPA), 1995d). Tankers and trucks can utilize vapor-balancing equipment, which sends VOC vapors displaced during filling back to the loading terminal for recovery (U.S. Environmental Protection Agency (U.S. EPA), 1995d). The selection of appropriate storage tank designs and routine maintenance can also mitigate emissions.

Volatile organic compound emissions can be controlled using either recapturing methods, or destructive methods. For low to medium concentration emission streams, VOCs can be adsorbed using a solid media such as activated carbon, silica gel, zeolite, or activated alumina (U.S. Environmental Protection Agency (U.S. EPA), 2002). Depending on the media and VOC properties, VOCs may be desorbed using heat, whereby regenerating the media. The most commonly used systems are fixed bed systems and canister systems. For high concentration (>5000 ppmv) emission streams, condensers can be used to recover VOCs (U.S. Environmental Protection Agency (U.S. EPA), 2002). Condensers are often used as prior to other control equipment in order to recover and recycle as great a volume as possible. Condensers are usually refrigerated to maximize recovery and can have removal efficiencies above 90% (U.S. Environmental Protection Agency (U.S. EPA), 2002).

Destructive methods for VOC emission control include flaring and incineration. Flaring is the open air burning of VOCs while incineration is the combustion of VOCs in an enclosed chamber. Both methods can provide excellent removal efficiencies (>98%), but require adequate mixing and oxygen levels to ensure complete combustion (U.S. Environmental Protection Agency (U.S. EPA), 2002).

Mitigation of Sulfur Dioxides Emissions

Mitigation of sulfur dioxide emissions should be implemented during all stages of metal processing where emissions are expected (e.g., roasting, smelting, sintering, copper converters, reverberatory furnaces, etc.). The most common method of mitigation is to channel the gases through on-site sulfuric acid plants (U.S. Environmental Protection Agency (U.S. EPA), 1995a,c,e). Sulfuric acid plants catalytically oxidize SO_2 into SO_3, which is then converted into sulfuric acid (H_2SO_4). Sulfuric acid can then be later used for electrolytic refining. These on-site plants can convert up to 99% of the SO_2 emissions into H_2SO_4, depending on their configuration (U.S. Environmental Protection Agency (U.S. EPA), 1995b).

Mitigation of vehicular SO_x emissions (and other air pollutants) was undertaken by the U.S. EPA through a two-pronged approach under the

Clear Air Nonroad Diesel Rule of 2004 (U.S. Environmental Protection Agency (U.S. EPA), 2004b). One approach was to require manufacturers to produce engines with advanced emission control technologies, such as selective catalytic reduction, in order to meet exhaust requirements (primarily NO_x). The other approach was to reduce the amount of sulfur in diesel fuel from 3000 ppm to 15 ppm by 2010, whereby lowering the potential SO_x emissions of all non-road vehicles running on diesel.

Mitigation of Nitrogen Oxide Emissions

The primary strategy for NO_x mitigation from blasting is to reduce the amount of toxic fumes produced. This can be achieved by using properly formulated explosives, avoiding deteriorated or contaminated explosives, ensuring proper confinement, avoiding reactions in diameters below the critical diameter, and using waterproof explosives in wet boreholes (Mainiero et al., 2007; Queensland Government, 2011). However, it is unlikely that complete mitigation will be possible and thus, effective management strategies such as wind monitoring and fume modeling should be implemented to ensure worker safety.

Mitigation of vehicular NO_x emissions (and other air pollutants) was undertaken by the U.S. EPA Clear Air Nonroad Diesel Rule of 2004 (U.S. Environmental Protection Agency (U.S. EPA), 2004b), which required manufacturers to produce engines with advanced emission control technologies, such as selective catalytic reduction, in order to meet exhaust requirements. Selective catalytic reduction systems reduce NO_x into nitrogen gas and water vapor and often run at removal efficiencies of 70–90% (U.S. Environmental Protection Agency (U.S. EPA), 2002).

Mitigation of Mercury Emissions

The use and atmospheric release of mercury during gold recovery is widespread in ASGM operations throughout the world. Organizations such as the Artisanal Gold Council, along with the U.S. Department of State and the United Nations are taking steps to introduce mercury control technology and train artisanal and small-scale gold miners on its proper use (Richard et al., 2014). Some of the largest obstacles to reducing mercury emissions continue to be proper education on the risks of mercury exposure, access to control equipment, and proper training.

One of the simplest mercury control devices is a retort. A retort consists of an airtight vessel in which the gold/mercury amalgam can be heated and the gaseous mercury retained and condensed back into liquid mercury. Recovery in this way both reduces the atmospheric emissions and recycles some of the mercury used in the amalgamation process. It is important to note that even when used properly, field retorts will reduce but not completely eliminate mercury emissions. Field retorts are expected to release 5–20% of the mercury vapor to the atmosphere (Richard et al., 2014). Retorts are a relatively simple technology and can be created using everyday items available to most ASGM operations, or bought pre-fabricated. Low cost retorts made of kitchen bowls and pipes have been shown to effectively reduce mercury vapor emissions up to 95% (Spiegel and Veiga, 2010). However, proper education and training on their use is still required, as some miners open retorts prematurely, releasing the accumulated vapors to the atmosphere.

In addition to retorts, the use of fume hoods with condensers or collection systems can be used to prevent exposure to mercury vapors and subsequent release into the atmosphere. Fume hoods alone, while preventing personal exposure to mercury vapors, will not prevent atmospheric emissions unless a condenser system is added. Argonne National Laboratory has demonstrated a mercury condenser that can be attached to existing fume hoods and observed an inlet-to-outlet reduction of mercury ranging from 60–80% depending on the design (Argonne National Laboratory, 2008). Another mercury condenser, the water based Kalimantan System, designed by the United Nations Industrial Development Organization's Global Mercury Project, uses a water condenser to reduce emissions by 75% (Argonne National Laboratory, 2008). Overall, the demonstration by Argonne National Laboratory concluded that any condenser systems be placed at a distance that allows six seconds of vapor transit time before reaching the condenser to allow mercury particles to coagulate to larger particles for more efficient removal, include baffles to increase removal, and that adherence of mercury to duct walls is an important contributor to mercury removal in such systems (Argonne National Laboratory, 2008).

Other technologies are being promoted for the reduction of overall mercury use in gold separation and concentration processes in ASGM. Conventional ore processing equipment such as sluices, shake tables, centrifuges, spiral concentrators, vortexes, and magnet separators are being encouraged for gold concentration as an alternative to whole ore amalgamation in order to reduce the amount of mercury required. However, implementation will increase the sophistication of the operation, requires significant capital, and increases labor requirements, all of which are significant obstacles for many ASGM operations in developing countries (Telmer and Veiga, 2008).

Mitigation of Methane Emissions

Because of the safety issues associated with the underground buildup of CMM, methane must be managed using dilution. Thus, maintaining an effective ventilation system is the easiest way to manage methane levels. Once diluted to safe levels and transported out of the mine, methane is vented to the atmosphere or collected for further use.

One method to avoid the release of CMM during mining is to pre-drain methane, called degasification, or to post-drain methane, called "gob" (Banks, 2012). Both systems require the drilling of boreholes to allow CMM to naturally vent out of the coal seam. Coal mine degasification operations can prevent atmospheric emissions of methane and can produce high quality methane. Degasification has also proven to be effective at reducing surface coal mine methane emissions (U.S. Environmental Protection Agency (U.S. EPA), 2014).

After collecting methane through a ventilation or degasification system, several control strategies exist for the treatment or removal of methane. If the recovered methane is of high enough quality to sell, methane may be injected directly into a natural gas pipeline or directly utilized on site for power generation or boiler fuel (United Nations, 2010). However, recovery of CMM is not economical if the methane produced is low quality or quantity. In this case, recovered methane can be flared and converted to carbon dioxide, reducing its global warming potential. In some areas, mines purify even low quality methane, resulting in a high-quality product that can be more widely utilized (United Nations, 2010).

One solution to manage large volumes of low concentration (typically below 1% methane by volume) CMM emissions from ventilation air is to use it as a combustion air source (as opposed to ambient air) for combustion turbines, internal combustion engines, or larger boiler or heating operations (U.S. Environmental Protection Agency (U.S. EPA), 2000). Doing so would utilize the low concentration of methane and also reduce fuel requirements for respective processes. Emerging technologies, such as thermal and catalytic flow-reversal reactors are capable of direct oxidation of low concentrations of methane, producing water and CO_2 as waste products (U.S. Environmental Protection Agency (U.S. EPA), 2000).

In non-coal mines with a history of methane emissions, methane recovery technologies similar to coal mines can be implemented to capture and use methane released from these mines and reduce emissions. Projects such as the Beatrix Gold Mine in South Africa and the Green River Trona Mine in Wyoming currently implement methane recovery systems for power generation, flaring, and fueling drying operations (U.S. Environmental Protection Agency (U.S. EPA), 2013).

Ventilation

Maintaining adequate underground mine ventilation is vital to prevent the build-up of hazardous gases, including methane and carbon monoxide, and maintain a safe work environment. In typical mine ventilation systems, air enters the mine via a shaft or other connection, flows along the intake pathway to the working areas picking up contaminants, returns through the system via return pathways, and finally is returned to the surface via shafts (Darling, 2011).

Both active and passive devices are used in ventilation systems to direct air along desired routes. Active devices, typically fans, are the primary method of producing airflow within the system. Surface fans are utilized as the primary airflow generators, as they allow for easier installation and maintenance (Darling, 2011). Booster fan systems can be placed throughout the mine, when necessary, to provide additional regulation of flow and pressure (Darling, 2011).

Passive devices, such as air doors, airlocks, regulators, air crossings, seals, and stoppings add resistance to airflow, whereby acting as a means to control the route of airflow (Darling, 2011). Air doors are used to prevent short-circuits in the flow of air, while still providing access between airflow intake and return routes. Airlocks are sets of two or more air doors, which are designed to be self-closing using air pressure. Passive regulators are adjustable orifices built into air doors. By changing the regulator opening, air flow through a certain area can be adjusted as required. Air crossings are areas where intake and return cross over one another, creating the possibility of leakage between the two (Darling, 2011). Air crossings typically consist of one airway elevated above the other, with a separating layer in between the two. This separating layer can consist of naturally occurring strata, masonry, concrete, steel structures, timber shuttering, or prefabricated fire-resistant structures (Darling, 2011). Seals are used to block off abandoned sections of the mine and prevent oxygen from mixing with potentially explosive gases within the section, but also reduce unwanted air loss in the ventilation system. Stoppings are used to block off access routes that are no longer necessary and reduce short-circuiting in the system (Darling, 2011).

7.3 MITIGATION OF WATER QUANTITY AND QUALITY IMPACTS

Mitigation of Water Quantity Impacts

Efforts should be taken to minimize impacts to surface water and groundwater quantity. The amount of water required for mining operations can be reduced using the adage "reduce, reuse, recycle" (Gunson et al., 2012). Many actions that can be taken to reduce water use rely on regular maintenance and upkeep of processing equipment, piping, fittings, water monitoring, and water storage. The mining industry is one of few that typically does not require water of high quality for many operations. As a result, water from sources that are not fit for human consumption (e.g., seawater, saline groundwater, treated wastewater from other operations) can be used for some operations, such as dust control, floatation, and other mineral processing operations (International Council on Mining & Metals (ICMM), 2012). Doing so greatly reduces the need to rely on potable surface water and groundwater. Two key concepts that can greatly improve mine water systems are to run processes at the highest solids densities possible and to use the lowest quality water that does not impact performance (Gunson et al., 2012).

The major sources of water loss during mining is from evaporation and seepage, water entrained in tailings, and tailings storage facilities (TSF) (Gunson et al., 2012). Methods for reducing water use are shown in Table 7.2. Not all methods are viable for every operation; water saving solutions should be tailored to each specific mining operation.

Water from tailings deposits and TSF ponds, along with effluent from tailings and concentrate thickeners, are commonly recycled as process water in mines throughout the world (Gunson et al., 2012). Oftentimes using recycled water for process water can reduce reagent requirements, but may result in the buildup of undesirable constituents, such as salt (Gunson et al., 2012). Such limitations may be overcome by freshwater blending or treating a small side stream to reduce constituents of concern. These methods can decrease both water demand and water treatment costs.

Through the implementation of water recycling and reduction processes, mining operations throughout the world have begun to reduce water use. Successful implementation of water recycling means that water consumption will no longer depend on water requirements for individual processes, but on permanent sinks such as evaporation, tailing or ore retention, and seepage (Gunson et al., 2012). For example, with the implementation of water recycling, water requirements for froth floatation in copper mines have been brought down to as low as 0.34 m^3 of water per ton of ore processed as compared to $1.9–3.0 \text{ m}^3$ per ton in operations without recycling (Gunson et al., 2012).

Table 7.2 Actions that can be implemented to reduce mine water use (Gunson et al., 2012)

Method	Explanation
Reduce	• Reducing wet area/open area in the tailings storage facility (TSF) • Reducing clay generation during grinding to lower tailings water retention • Improving tailings thickener performance • Reducing water losses through thickened tailings or paste tailings disposal • Reducing water losses through the installation of drains in the TSF • Reducing water losses through tailings compaction • Reducing water losses through selective tailings size classification • Reducing water losses by using tailings filtration • Reducing the concentrate moisture content • Reducing evaporation through covers (tanks/thickeners/tailings pond) • Reducing evaporation through alternative dust suppressants • Ensuring no unplanned pipeline water losses • Eliminating evaporative cooling • Reducing pump gland seal water usage • Reducing site employee/contractor water use • Reducing water consumption through ore pre-concentration • Reducing water use through dry processing
Reuse	• Collecting and reusing rainwater and surface runoff water • Reusing mine dewatering water • Reusing offsite waste water • Reusing cooling water • Reusing grey water
Recycle	• Recycling TSF surface water • Recycling TSF seepage water • Recycling tailings thickener overflow • Recycling concentrate or intermediate thickener overflow • Recycling potential mine effluent water

Each action may not be applicable for all mining operations.

Water Quality Control
Mitigation of Water Quality Impacts

During mining, the generation of contaminated water with the potential to negatively impact water resources is inevitable. Thus, methods to mitigate water quality impacts primarily focus on reducing the amount of contaminated water generated and preventing it from contacting surface and groundwater (International Council on Mining & Metals (ICMM),

2012). The generation of some wastewaters may only occur during specific phases of a mining operation; however, some may persist long after the mine is closed (e.g., acid mine drainage, mine dewatering) and require an ongoing management plan to prevent water quality impacts.

Wastewater from mining operations and runoff tends to be highly acidic and cannot be discharged without proper treatment. Depending on the pH, contaminants present, volume of wastewater, and discharge requirements, various treatment technologies can be implemented. Highly acidic wastewater typically requires the addition of lime or caustic soda to increase the pH. Increasing the pH also tends to precipitate out dissolved metals, which can be further removed by adding coagulants and flocculants followed by settling. More advanced treatment technologies such as reverse osmosis, membrane filtration, and ion exchange resins may be used to further increase water quality as required by discharge permits. Similar methods can be used to treat contaminated groundwater during pump and treat operations. Wastewater generated from ore processing should be treated to meet desired water quality requirements and recycled to reduce water requirements, wastewater generation, and the quantity of processing chemicals used. In some environments, wastewater volume can be reduced before treatment by using evaporation ponds.

In addition to conventional water treatment technologies, other technologies and methods have been developed to address specific constituents of concern (e.g., heavy metals) in mine wastewater. Select technologies for addressing heavy metal contamination are listed in Table 7.3

Risks of groundwater contamination due to AMD, water from tailing ponds, or leachate from waste rock piles can be controlled by draining and capping tailings ponds, and lining and covering waste rock and ore piles (Butler, 2014). Proper capping with an impermeable layer will prevent precipitation from infiltrating tailings and reduce the need for leachate removal and treatment (Butler, 2014). Additional rehabilitation steps, including covering piles with topsoil and revegetation, can help restore the area closer to its natural state and prevent erosion and breakthrough of the impermeable layer (Butler, 2014). Tailings from heap leaching operations should be rinsed to remove residual heap leaching solution before being capped.

Further groundwater contamination can be avoided by directing runoff away from areas that have been filled and capped. Rain and snowmelt flowing through the site should be captured using a drainage system and channeled to tailings ponds to prevent the transport of contaminants offsite. Contamination of groundwater can be controlled by injecting grout curtains, whereby

Table 7.3 Technologies for treating heavy metals in mine wastewater
(U.S. Environmental Protection Agency (U.S. EPA), 2004a)

Technology	Description
Arsenic oxidation	Photochemical oxidation of arsenic to arsenate, and subsequent removal via iron co-precipitation
Bauxsol	A mixture of iron and aluminum oxyhydroxides and alumino-hydroxy-carbonates that sequesters heavy metals and causes them to quickly precipitate out of solution
Biological reduction of selenium	Proprietary microorganisms in biofilms in anaerobic solids bed reactors precipitate elemental selenium with 97% reduction
Catalyzed cementation	Removal of heavy metals (primarily selenium) by catalytic cementation on the surface of iron particles
Colloid polishing filter method	Removal of ionic, colloidal, radionuclides, and heavy metals by sorption and chemical complexing
Ionic state modification process	A proprietary chemical is mixed in before applying a strong magnetic field to the wastewater. Heavy metals are oxidized or reduced to where they can be precipitated out using conventional technology
Silica micro encapsulation	Encapsulation of heavy metals in a silica matrix to prevent migration or environmental interactions
Redox-mediated biotransformation	Removal of heavy metals, nitrate, and sulfate using in-situ biological treatment
Alumina adsorption of arsenic	Alumina is used to adsorb arsenic at low pH before being filtered out and regenerated using sodium hydroxide to remove arsenic

diverting groundwater and preventing it from moving through areas that may cause contamination (U.S. Environmental Protection Agency (U.S. EPA), 2004a).

Mitigation techniques will vary depending on site-specific factors. For example, mining operations in areas with high rates of precipitation should focus on capturing or reducing water flowing off site, while operations in arid environments should focus on water treatment and recycling (International Council on Mining & Metals (ICMM), 2012).

Acid Mine Drainage
Mitigation of Acid Mine Drainage

When dealing with AMD, prevention and mitigation are generally preferable to treatment. Minimizing AMD reduces harmful environmental effects and the excessive lifetime treatment costs associated with AMD

Table 7.4 Acid mine drainage mitigation techniques (Jennings et al., 2008; Johnson and Hallberg, 2005; Leading Practice Sustainable Development Program for the Mining Industry, 2007b; U.S. Environmental Protection Agency (U.S. EPA), 2004a)

Mitigation approaches	Comments
Avoid sulfide deposits	May be difficult as sulfidic minerals can be associated with the ore of interest and soils are not always homogenous. Careful mine planning may avoid sulfidic deposits. Underground mining versus surface mining can potentially reduce risk.
Selective handling of AMD forming materials	Inherent difficulty to separate acid-forming materials from benign overburden. Encapsulate or mixing acid forming wastes with benign or buffering waste materials. Backfilling mines with tailings, Portland cement, and other minders can neutralize potential AMD.
Avoid oxidizing conditions	Mine flooding or subaqueous disposal of tailings and waste rock below the water table prevents contact with oxygen. Tailings can be flooded under shallow water covered by sediments.
Avoid water contact	Top soils and clay sealing layers need to be resistant to breakthrough by erosion, plant roots, and burrowing animals. Proper design should mimic natural slopes to limit sediment loss and incorporate other erosion protection strategies such as vegetation and rip rap. These need to be maintained in dry climates to prevent cracking and erosion. Clean water and runoff should be diverted away from the waste storage sites.
Microorganism inhibition	Applying biocides can inhibit microbial activity and reduce AMD. Generally a short-term solution requiring repeated applications.

issues. Mitigation approaches for AMD are listed in Table 7.4. Mitigation measures typically act by removing either water or oxygen (or both) from the system, preventing sulfide oxidation (Johnson and Hallberg, 2005). These are most effective when used in conjunction with each other. Applicability must be determined on a site specific basis.

Mine flooding due to runoff or groundwater after operations cease is always a possibility. Water that enters the mine can dissolve exposed ore and waste rock, forming AMD. This can be avoided by sealing the mine or intercepting and redirecting groundwater flowpaths to avoid mined areas (Butler, 2014). Dry mine seals, consisting of concrete or masonry walls, can be used to keep water out of the mine while wet mine seals, consisting of concrete blocks with holes or pipes for drainage, allow water to flow into the mine but prevent air from entering (U.S. Environmental Protection Agency (U.S. EPA), 2004a). Once mine flooding becomes a problem, the mine can be completely inundated and sealed to prevent AMD by limiting the supply of oxygen. Initial dissolved oxygen in the water will be consumed by microorganisms and the diffusion of additional oxygen will be prevented by sealing the mine. Care should be taken to prevent oxygen from entering the mine, as the formation of AMD is likely to contaminate the surrounding groundwater. Water can also be continually pumped out of the mine to prevent AMD. Water that is pumped out of mines in de-watering efforts can be treated and reused for other purposes.

Surface contact between sulfide ores and water can be minimized by sealing spoils, tailings, and waste rock under a layer of clay or another hermetic material. These piles should also be on an impermeable base to prevent any leachate from contaminating groundwater (Johnson and Hallberg, 2005). Additionally, runoff should be channeled away from these key areas to avoid water contact and subsequent AMD. Alternatively, mine tailings can be covered in shallow water and a layer to sediments to limit contact with dissolved oxygen in a method similar to mine flooding (Johnson and Hallberg, 2005).

While AMD prediction and mitigation techniques are costly (Jennings et al., 2008), insufficient evaluation and prevention can result in negative environmental impacts and even higher treatment costs. For example, treatment costs for AMD at Iron Mountain Mine, California, are estimated at around $950 million and rising (U.S. Environmental Protection Agency (U.S. EPA), 2006).

Acid Mine Drainage Control

If mitigation of AMD fails or cannot be implemented, control and treatment is the next alternative. Treatment of AMD can be classified as either passive or active (Table 7.5). Active treatments typically refer to the ongoing treatment of AMD using water treatment facilities. Active treatment operations employ technology similar to conventional wastewater treatment with the exception of utilizing chemical agents and aeration to neutralize acidity and precipitate out metals (Johnson and Hallberg, 2005). Active technologies for AMD treatment are the most common and chemicals used vary depending the cost, effectiveness, and effluent requirements. Common neutralizing chemicals include lime, calcium carbonate, sodium carbonate,

Table 7.5 Types of treatment used for AMD control (Jennings et al., 2008; Johnson and Hallberg, 2005; Leading Practice Sustainable Development Program for the Mining Industry, 2007b; U.S. Environmental Protection Agency (U.S. EPA), 2004a)

Type of treatment	Description	Examples	How to choose treatment
Passive treatment	No chemical amendments, no motorized/ mechanized assistance	Wetlands, limestone drains and ponds, alkalinity producing systems, permeable reactive barriers, packed bed iron-oxidation bioreactors	Choose passive treatment for situations with low acidity loads and steady flow rates
Active treatment	Highly engineered water treatment facilities that can accommodate a wide range of pH levels	Precipitation of metal hydroxides, coagulation and flocculation, ion exchange, membrane separation, bioreactor systems	Effective for variable flow rates and pH. Treatment technologies depend on costs, dosage rate, extent of preparation and delivery system needed, sludge characteristics, and discharge requirements

sodium hydroxide, or magnesium hydroxide (Johnson and Hallberg, 2005). In addition to traditional treatment plants, more efficient and less expensive treatment alternatives have been developed. For example, the in-line aeration and neutralization system developed by the U.S. Bureau of Mines mixes sodium hydroxide or sodium carbonate and aerates AMD using Venturi action and a static mixer before discharging into a clarifier, in what is essentially a pipeline version of a conventional treatment plant (U.S. Environmental Protection Agency (U.S. EPA), 2004a).

Passive treatments, such as constructed wetlands and limestone drainage systems, require relatively little ongoing maintenance when compared to active treatments, but may be impractical or expensive set up (Johnson and Hallberg, 2005). While passive treatment can be more cost effective in some cases, it has to be chosen and designed very carefully avoid system failure (Leading Practice Sustainable Development Program for the Mining

Industry, 2007b). For example, anoxic limestone drainage systems act by increasing the alkalinity of AMD and raising the pH. This causes metal hydroxides to precipitate out of solution, which can decrease drain permeability and performance (U.S. Environmental Protection Agency (U.S. EPA), 2004a). In some cases, limestone drains can be modified to accumulate and recover precipitated metals such as aluminum (U.S. Environmental Protection Agency (U.S. EPA), 2004a). Multiple passive systems used in series can increase overall treatment and reduce individual maintenance. For example, alkalinity producing systems, where AMD filters through organic materials to consume oxygen and precipitate iron out of solution, can be utilized before limestone drains to reduce drain clogging (U.S. Environmental Protection Agency (U.S. EPA), 2004a). Other passive treatment such as permeable reactive barriers can be used to treat contaminated groundwater.

7.4 MITIGATION OF LAND IMPACTS
Topsoil Management

In order to return the land back to its original condition during rehabilitation, topsoil must be managed throughout the entire life of a mine. The primary goals of topsoil management are to maintain sufficient water retention to support natural vegetation through dry seasons, sufficient depth for root growth, and nutrient cycling capacity (Leading Practice Sustainable Development Program for the Mining Industry, 2007a). During planning stages, physical properties of the topsoil must be collected, including (Leading Practice Sustainable Development Program for the Mining Industry, 2006):

• Depth of topsoil
• Volume of topsoil to be collected and types of equipment needed
• Erodibility
• Re-spreading depth
• Re-spreading follow-up treatment

Knowing these attributes will assist mine operators in making topsoil management decisions during subsequent stages of the mine's life.

During construction of the mine, soil should be stripped and stockpiled in separate phases. Because soil found deeper in the profile may not have the same nutrients or seed content found in the topsoil (most seeds are concentrated in the top 50 mm), it is important to avoid mixing of topsoil and deeper soil (Leading Practice Sustainable Development Program for the Mining Industry, 2006). Top soil should be stripped and collected during the dry season if possible. Stockpiling of topsoil should be avoided as much as possible; however, when necessary, stockpiles should be kept at lower heights (1–3 meters) and should have a life span shorter than 12 months (Leading Practice Sustainable Development Program for the Mining Industry, 2006). During stockpiling, topsoil can be seeded with a grass/legume mix to reduce erosion and the loss of beneficial soil microorganisms. Early rehabilitation should be performed on open sites that are no longer needed to reduce stockpiling time. When topsoil is re-spread during rehabilitation phases, soil treatments can be used if native topsoils are dispersive (soils that become denser when wet, restricting root growth) or have high salinity. It can be necessary to inoculate soil with microorganisms or to fertilize the soil as stockpiling the soil tends to have a negative impact on microbial activity and nutrient availability (Harris et al., 1987; Mummey et al., 2002; Stahl et al., 1988).

Mitigation of Heavy Metal Contamination

Proper management of heavy metal contaminated soil, for example controlling soil pH, draining wet soils, and capping waste piles, can be viable alternatives to minimize the risk of transportation offsite, reduce bioavailability, and minimize future exposure (U.S. Department of Agriculture, 2000). Heavy metals can be prevented from solubilizing and migrating offsite by maintain a soil pH of 6.5 or higher (U.S. Department of Agriculture, 2000). Thus, proper mitigative and control measures should be taken to prevent acid mine drainage from coming into contact with contaminated soils. Wet soils should be drained in order to promote aeration and subsequent oxidation of metals into less soluble forms (U.S. Department of Agriculture, 2000). Dust control technology should be implemented to prevent wind erosion and the dispersion of heavy metal containing dust throughout the site by vehicle traffic. Emission control, including scrubbers and condensers can prevent heavy metal particulates from being released into the atmosphere and being deposited on the soil. Tailings and waste rock piles should be properly managed; steps should be taken to ensure waste rock is not used for outside construction projects, tailings dams should be inspected on a regular basis to prevent failure, and tailings and waste rock should be capped with impermeable barriers to prevent leaching. Wastewater and other mine processing wastes should be treated and properly disposed of. Because heavy metal remediation options are limited, it is best to prevent heavy metal contamination from occurring through proper management (Li et al., 2014).

Many available heavy metal remediation options are expensive, lengthy, and may not always restore soil productivity (U.S. Department of Agriculture, 2000; Wuana and Okieimen, 2011). One method to remediate heavy metal contaminated soils is to use plants to extract or immobilize the heavy metals via phytoremediation. Some species of plants naturally uptake large amounts of heavy metals and will translocate the metals to their stems and leaves in a process called phytoextraction (U.S. Department of Agriculture, 2000). These plants can then be harvested and properly disposed of. Other species of plants will uptake heavy metals into their root systems or can cause metals to precipitate in the rhizosphere in a process called phytostabilization (U.S. Department of Agriculture, 2000). Although this does not remove heavy metals, phytostabilization prevents the further movement of heavy metals via erosion and leaching, and reduces overall bioavailability (U.S. Department of Agriculture, 2000). In addition to plants, chemical amendment can be used to increase or decrease the mobility of heavy metals in soils. Chemical amendments that increase heavy metal solubility (e.g., chelating, complexing, and desorption agents) can be used in conjunction with soil washing and phytoextraction, but comes at the risk of increased heavy metal leaching and possible groundwater contamination (Bolan et al., 2014). Immobilizing chemical agents, such as phosphate compounds, organic matter, and liming agents, can be used in conjunction with phytostabilization to further reduce mobilization and bioavailability (Bolan et al., 2014).

Slope Stabilization

When slope failure is likely, the simplest solution is to evacuate the area and let the material fall (Girard, 2001). This is a good option for non-vital areas, but may not be feasible for active mining areas. In these cases, slope stabilization is desirable. Slope stabilization methods consist of three primary categories: reinforcement, rock removal, and protection (Wyllie and Mah, 2004) (Table 7.6). If initial analysis indicates a high chance of slope failure, the slope can be reinforced using fully grouted untensioned dowels and tensioned rock bolts (Wyllie and Mah, 2004). Rock bolts increase friction along fracture planes between the unstable surface and the stable interior and can be either strength bolts, yieldable bolts, or energy-absorbing bolts (He et al., 2014). Furthermore, slopes have been successfully reinforced and stabilized using combinations of bolts, cables, meshes, shotcrete, and buttresses (Girard, 2001). Meshes installed on the slope face can prevent rockfall, and ditches at the toe of the slope can be

Table 7.6 Measures for slope stabilization (Wyllie and Mah, 2004)

Slope stabilization category	Stabilization measure
Reinforcement	• Rock bolting • Dowels • Shotcrete • Drainage • Buttresses • Shot-in-place buttress • Tied-back walls
Protection measures	• Ditches • Mesh • Catch and warning fences • Rock sheds • Tunnels
Rock removal	• Resloping • Trimming • Scaling

used to catch rockfall. Shotcrete can be applied directly to the rock face to protect against rock fall and provide surface stabilization, but does little to prevent sliding of the slope (Wyllie and Mah, 2004). Overhangs and cavities that form on the slope can be stabilized using concrete buttresses to support the overhanging layers. These reinforcement options can be very expensive and maintenance requirements should be considered when determining the most cost-effective approach to slope stabilization.

Resloping, or flattening the angle of an unstable slope, can remove the driving force of slope failure and eliminating the hazard (Girard, 2001). A small bench can be left at the base of the resloped area to act as a small catchment area and to provide a base for future resloping operations (Wyllie and Mah, 2004). Trim blasting of potentially hazardous overhanging rock and hand scaling of loose rock and soil can provide further stabilization; however, care should be taken to prevent the undermining of erodible soil matrix areas (Wyllie and Mah, 2004).

Subsidence Mitigation

Subsidence can be reduced through proper mitigation techniques including altering mining methods, planning and monitoring, and post-mining stabilization (Blodgett and Kuipers, 2002). Most subsidence mitigation measures aim to prevent damage to infrastructure and buildings, rather than negative impacts to wildlife and water quantity and quality.

The severity of subsidence can depend on the method of underground mining. Alterations to mining methods including extraction rates, mine layout and configuration, partially mining to leave protective pillars, and backfilling can reduce, but not entirely eliminate, subsidence (Blodgett and Kuipers, 2002). During mine planning, the amount of subsidence should be taken into account so all buildings and infrastructure can be properly designed to withstand differential settling caused by the estimated subsidence. This can be done through modeling using analytical methods.

Subsidence can be reduced through the backfilling of mines, which provides support to the ground in direct contact with the backfill. Backfilling can be widely categorized as either hydraulic, paste, or rock backfill (Blodgett and Kuipers, 2002; Yao et al., 2012). Hydraulic backfilling involves forming a slurry consisting of water, sand, aggregate materials, and a binder to backfill a mine. Paste backfills differ from hydraulic backfilling in that the mixture contains a larger fraction of silt sized particles and lower water content. Rock backfilling primarily consists of a mixture of coarse aggregate, sand, and possibly tailings to achieve desired grain size distributions. Backfilling can utilize a variety of materials, including mine tailings, overburden, sand, course aggregate, Portland cement, blast furnace slag, and fly ash (Blodgett and Kuipers, 2002; Siriwardane et al., 2003; Sui et al., 2014; Yao et al., 2012). Not only does backfilling increase roof support, it can also serve as a method of waste rock and tailings disposal and in the process reduce water quality impacts such as acid mine drainage (Blodgett and Kuipers, 2002).

7.5 MITIGATION OF ECOLOGICAL IMPACTS

It is important that ecological impacts be considered during the planning phases of mining projects, as biodiversity and habitat, once lost, are difficult to fully recover. Mine designers and operators can reduce ecological impacts through baseline monitoring and incorporating mitigation measures during initial stages of mine planning. During operational stages, integrated biodiversity management and monitoring can be used in conjunction with environmental management systems to avoid, reduce, remedy, and compensate ecological impacts. Baseline monitoring allows mining companies to identify the existing biodiversity, biodiversity values of the indigenous communities, and key risks to biodiversity. Using this data, mitigation techniques can be incorporated into mine planning and design. Furthermore, proper control and mitigation of negative air quality, water quality, and land impacts can also greatly reduce negative ecological impacts.

Mine Planning

After risk levels for ecological impacts have been determined using baseline monitoring, different management styles can be used depending on the existing ecological condition and the proposed mining project. These management styles are presented in Table 7.7. Holistic land management, the first management style addressed in Table 7.7, should guide the decision making process for choosing mitigation techniques. Ideally, mine layouts and operations should be planned in a way to minimize land and water disturbances to sensitive habitats. Access roads, railways, and utility lines should avoid critical habitats and limit habitat fragmentation. When fragmentation is unavoidable, efforts should be taken to group fragments close together, rather than having fragments dispersed throughout an area. Land clearing, building layouts, and waste disposal methods should be designed to limit habitat destruction and possible contamination to surrounding areas. Laws such as the Stream Buffer Zone Rule prevent, whenever possible, mining activities on land within 100 ft of perennial or intermittent streams and require that such mining activities would not cause the stream to violate Clean Water Act standards (Copeland, 2014). Other legislation, such as the Endangered Species Act, is meant to protect critical habitats from disturbances and pollution caused by mining and other activities.

Conservation Banking and Offsetting

Unfortunately, mine planning can never completely eliminate negative ecological impacts. When impacts are unavoidable, methods such as conservation banking may be implemented. Conservation banking is a form of offsite mitigation that creates a permanently protected land for listed species habitat use in exchange for unavoidable impacts on a species at a project site (U.S. Fish & Wildlife Service, 2012). These types of offsetting measures should ideally preserve "like-for-like" habitat, or habitat that is ecologically equivalent to the one that will be destroyed, although preserving habitats that have greater biodiversity or ecological importance may be acceptable as well (Virah-Sawmy et al., 2014). One advantage to this approach is that after mining operations are completed, the project site can be rehabilitated and the overall species habitat may increase as a result of the conservation bank. Many companies, such as Rio Tinto and major investors have begun

Table 7.7 General mitigation styles that can be used individually or in conjunction with one another to decrease ecological impacts around mining areas

Biodiversity management style	Why technique is needed	Application of management techniques
Holistic land management	Ecological effects overlap manmade boundaries and impacts on a mine site could translate to impacts at a site not touched by the mine.	Mine owners should communicate with nearby landowners and community groups to define conservation values, forge partnerships with government, identify pathways species use to move across landscapes, and prevent colonization in mining areas.
Initiate and maintain ecological services	Mining can involve a high level of disturbance of land so maintaining habitat availability is essential to maintaining biodiversity.	Maintenance of habitats for threatened species, water quality upkeep, and natural pest control should continue until the mine has been rehabilitated. Incorporate progressive rehabilitation programs during mine life.
Biodiversity offsets	If mining actions create unavoidable harm to biodiversity, offsets can be used to compensate for harm. Offsets are a controversial mitigation technique as it is difficult to measure if implementation is effective.	Rehabilitate and/or re-establish degraded ecosystems. Acquire and include native vegetation into rehabilitated site if it is being threatened. Fence remnant vegetation. Contribute to knowledge through monitoring and research. Provide resources to local conservation or land care groups. Prepare and implement agreed recovery plans for specific species.
Conservation banking	A subset of biodiversity offsets, this method provides offsets during mining, versus having to wait for mine rehabilitation.	Resource conservation sites can be chosen and managed to maintain/improve natural resource value. The chosen sites do not need to be in the mine area.

Adapted from Leading Practice Sustainable Development Program for the Mining Industry (2007a).

to incorporate "no-net-loss" and "net positive impact" principals for biodiversity into their policies for both ecosystem protection and to boost their public image (Virah-Sawmy et al., 2014).

One of the key issues facing banking and biodiversity offsetting is that biodiversity is difficult to quantify, and thus, the effectiveness of offsets cannot be easily measured nor can "net positive impacts" be effectively determined. The most commonly used biodiversity measurement is expressed as (Virah-Sawmy et al., 2014):

Biodiversity Net Change

$$= (A_{gain})(Biodiversity\ value\ at\ offset\ site)$$
$$- (A_{loss})(Biodiversity\ value\ at\ impacted\ site)$$

where:

A_{gain}, A_{loss} = habitat area gain and loss, respectively; hectares.

Calculating the biodiversity value, while simple and cost effective, is problematic, as it cannot account for rare or threatened species, unique ecological communities, species vulnerability, or habitat requirements for endangered species (Virah-Sawmy et al., 2014). Biodiversity calculations may only take into account one factor, such as plant diversity or bird diversity, which can cause the habitat gain or loss factors to vary significantly.

Another issue with banking and biodiversity offsetting is how "no-net-loss" and "net positive impact" are calculated. For example, some companies may determine that if mining activities did not take place at a given location, the land would be implemented for agricultural use instead. Thus, the mining company could claim that there is no net loss of habitat, because the habitat would have been destroyed regardless. The lack of established standards for how impacts to biodiversity and habitat are calculated hurts the effectiveness of a potentially beneficial program.

Mine Operation

After the general mitigation style has been chosen and the mine is in operation, the day-to-day operations can be managed to avoid and reduce impacts. Table 7.8 provides examples of mitigation techniques that can be used to reduce impacts to both flora and fauna. As these mitigation techniques should be put into practice during mine operation, they must first be conceptualized during the mine planning stage and incorporated into the environmental management plan to maintain accountability.

Table 7.8 Specific mitigation measures that can be incorporated into the mine plan for both flora and fauna are listed below (Leading Practice Sustainable Development Program for the Mining Industry, 2007a)

Impacted ecological features	How to minimize impact on ecological feature
Terrestrial vegetation and fauna	Survey area, analyze data, develop plan to protect areas
	Minimize clearing where high values of flora and fauna are located
	Set up appropriate fire regimes
	Land management plans should include:
	Identify and control of weed problems
	Identify, monitor, and control (if necessary) feral fauna that negatively impact other species and control non-native species
	Tailings storage areas should be made unattractive to fauna
	Consider connectivity of vegetation on site is connected to vegetation off-site (for animal movement)
	Identify favored animal crossing points of roads and install proper signage to reduce wildlife impacts
Aquatic fauna	Develop a water quality risk management framework to decrease impacts on biodiversity and use appropriate trigger values to define quality and quantity problems
	Understand how groundwater quantity affects surface water quantity
	Consider how existing flows and pathways will be altered during and after mining
	Manage and monitor wastes that could potentially contaminate water bodies
	Create compensating habitat structures for areas that are highly affected
Aquatic, riparian, and groundwater dependent vegetation	Consider potential hydrological or water quality changes that could impact vegetation
	Predicting changes in water quality and quantity and how the changes impact biodiversity can be difficult
	Develop monitoring and research programs to understand impacts
	Develop management programs to deal with impacts

In addition to the specific mitigation measures that can be implemented to reduce impacts on specific ecological features, best management practices can be implemented to reduce all aspects of ecological impacts. If water resources systems are properly managed through the careful consideration of quality, quantity, and timing, ecological risks and impacts should be inherently reduced as well. However, in order to implement standards for quantity, quality, and timing, it is necessary to know what living standards dependent flora and fauna require. In general, best management practices require "good housekeeping," a practice that encourages proper site maintenance and adherence to all safety and materials handling guidelines.

7.6 FIRE AND EXPLOSION CONTROL

Fires and explosions are a continuous risk throughout the life of a mining operation. Coal fires can occur due to the spontaneous exothermic reaction between coal and oxygen, where coal can reach the ignition temperature if heat is not dissipated (Lu et al., 2015). Explosions can occur due to the ignition of coal dust, methane, other flammable gases such as acetylene or hydrogen, or any combination of the three (Mine Safety and Health Administration (MSHA), 2006b). Coal dust explosions tend to be the most catastrophic of the three.

Fires and explosions require fuel, heat, and oxygen to occur. Typical fire prevention measures focus on removing the fuel or heat from potentially hazardous situations. Primary preventative measures include (Mine Safety and Health Administration (MSHA), 2006b):

- Frequent testing for methane and other flammable gases
- Maintaining adequate ventilation and abandoned area seals
- Maintaining water sprays and dust collectors to control coal dust
- Keeping mine surfaces wet
- Proper application of rock dust
- Reduce contact of coal with oxygen
- Using extra precaution when mining in potentially high risk areas
- Cleaning up loose coal, coal dust, and flammable or combustible materials

The frequent testing for methane and other flammable gases is important to prevent fires and explosions. Abandoned areas, retreat mining sections, and other areas that only require periodic work are especially important to monitor, as the build-up of methane can go undetected until it is too late (Mine Safety and Health Administration (MSHA), 2006b).

Maintaining adequate ventilation systems and seals on abandoned areas is vital to reduce the buildup of flammable gases in the atmosphere.

However, these systems are vulnerable to changes in atmospheric pressure (Mine Safety and Health Administration (MSHA), 2006b). The sealing of abandoned areas reduces air loss from ventilation systems and prevents oxygen, which may lead to spontaneous combustion, from entering these areas. Although seals for areas containing non-inert atmospheres must be designed to withstand an overpressure of 120 psi, they should be constantly monitored for outgassing or ingassing of methane or other hazardous gases.

Water sprays and dust collectors along the working face reduce the amount of coal dust in the air at the point of generation (Mine Safety and Health Administration (MSHA), 2006b). This also helps to reduce the dispersion of small coal dust particles further away from the working face. Additionally, water sprays also act to keep mine surfaces wet, preventing any settled coal dust from being re-fluidized (Mine Safety and Health Administration (MSHA), 2006b). Further away from the working face, water is less effective for controlling coal dust. Water will readily evaporate from coal dust depending on changes in the weather or ventilation system, turning safe areas into hazards in a short period of time (Harris et al., 2010). As a result, many mine explosions occur in the winter, when humidity is generally low and moisture content of dust is variable. For areas away from the working face, rock dust, for which drying and evaporation are not factors, can be applied.

Other fire prevention and control methods to reduce coal contact with oxygen include inert gas injection, grouting, applying chemical agents, and foaming agents. Although inert gas injection is effective, gases are prone to diffusion and may not remain close to the injection point unless areas are adequately sealed (Lu et al., 2015). Chemical agents such as sodium silicate and ammonium salts can be expensive and the range of application can be limited, especially during fire control situations (Lu et al., 2015). Grouting slurry can consist of fluidized bed combustion ash, water, and other ingredients. Grouting slurry is effective at managing fires but has difficulty managing the upper areas of a fire due to gravity and settling (Lu et al., 2015).

Rock Dust

The use of rock dust, made simply of finely crushed rocks such as limestone, dolomite, anhydrite, shale, adobe, or gypsum, is required by law in coal mines to reduce the risk of coal dust explosions (Mine Safety and Health Administration (MSHA), 2006b). In coal dust explosions, the generated air pressure and turbulence can fluidize and disperse surrounding coal dust, causing a propagating explosion (Harris et al., 2010). By covering coal dust with a layer of rock dust, the underlying coal dust is rendered inert and the rock dust acts as a heat sink, reducing the risk of propagating explosions. In the United States, all areas of a coal mine to within 40 ft of working faces must be adequately rock dusted at all times, as fine coal dust particles can travel a significant distance from the working face before settling (Mine Safety and Health Administration (MSHA), 2006b).

Rock dust is required to be distributed along the floor, sides, and top of all underground mine areas. The ceiling of underground mine areas is usually covered with various mesh, ribs, fencing, wood planks, or other materials to prevent roof skin failure. These areas can also collect coal dust and must be covered by an adequate amount of rock dust (Harris et al., 2010). Rock dust is dispersed via spray application to ensure even coverage. Current regulations by the MSHA require that 100% of rock dust must pass through a 20 mesh sieve (0.84 mm), 70% must pass through a 200 mesh sieve (0.074 mm), and that the incombustible content (rock dust) of the mixed dust layer to be at least 80% in return and non-return airway areas (Harris et al., 2010; Sapko et al., 2007; U.S. Department of Labor, 2010). Smaller particles of rock dust are more effective at preventing explosion propagation due to the greater surface area for heat absorption (Harris et al., 2010). The higher percentage of incombustible content is due to the accumulation of fine coal dust and the resulting higher explosivity (Harris et al., 2010). Furthermore, the incombustible content must be increased by 0.4% for each 0.1% of methane present in any ventilating current (U.S. Department of Labor, 2010).

The ratio of coal dust to rock dust can be monitored through the use of a coal dust explosibility meter, a handheld portable device developed by the National Institute of Occupational Safety and Health, which provides instant measurements (National Institute for Occupational Safety and Health (NIOSH), 2011). These handheld devices can also be outfitted to monitor methane in real time.

Foams

Foams are another alternative to control and prevent fires in underground mines. In addition to being used for coal dust suppression, foams can be injected into high fire risk areas and act to prevent fires by reducing contact between oxygen and coal by filling the fire zone (Qin et al., 2005). In longwall mining, the cavity behind the longwall where material has already been excavated and removed, known as the goaf, can contain residual coal. In longwall top coal caving mining, where roof coal materials in the goaf are allowed to collapse behind the working face, the goaf coal can spontaneously ignite or release gas and flood underground workings and fractures, increasing the risk of explosion or fire (Qin et al., 2005). Foams can be injected or sprayed directly into a goaf fire or can be pumped from above ground.

Foams can be effectively applied to prevent and control fire due to their high apparent viscosity, larger interfacial area, low filtrate and circulation losses, low cost, easy transportation, low density, and expansion properties (Lu et al., 2015). Water based foams, in addition to excluding oxygen, act to cool the coal through evaporation and may have further fire prevention properties if they are enriched with nitrogen gas to further displace oxygen (Trevits et al., 2007). Three-phase foams, consisting of a noncombustible solid phase (e.g., mud or fly ash), water, nitrogen gas, and a surfactant, have been used extensively throughout China to prevent and extinguish fires in underground coal mines (Qin et al., 2005). The solid phase increases the foam stability and can uniformly cover coal surfaces as the foam ruptures to prevent further contact with oxygen (Zhou et al., 2006).

Other foams, such as rigid polyurethane foams, phenolic foams, and urea-formaldehyde foams, are commonly used throughout China for fire control by managing air leakage (Hu et al., 2014). These foams have the added benefits of being able to stabilize and fill caving areas and seal off closed walls to prevent further air leakage and improve structural stability (Hu et al., 2014). However, some formulations may be flammable, toxic, or have relatively low compressive strength (Hu et al., 2014).

7.7 HUMAN HEALTH AND SAFETY

In 2006, The Mine Improvement and New Emergency Response Act, known as the MINER Act was passed and signed into law. This act contains provisions aimed to improve worker health and safety in mines throughout the United States. In addition to general safety provisions, the MINER Act requires the installation of refuge alternatives (refuge chambers and stations), post-accident communication systems and electronic tracking of trapped workers. Similar acts regulating mine safety and working conditions have been passed into law in other countries (e.g., Western Australia's Mine Safety and Inspection Act 1994, Law of the People's Republic of China on Safety in Mines, Canada's Labor Code Part II and

Coal Mining Occupational Health and Safety Regulations, South Africa's Mine Health and Safety Act 1996, etc.).

Personal Protective Equipment

During mining operations, personal protective equipment (PPE) is often required to protect against respiratory, hearing, and physical hazards. Typical hazards associated with mining may require the use of hard hats, protective footwear, hearing protection, safety glasses, face shields, safety belts and lanyards, gloves, chemical resistant clothing, or aprons (Mine Safety and Health Administration (MSHA), 2006a). However, mining poses a variety of unique hazards that call for specialized protective equipment.

Respirators are often used to prevent the inhalation of hazardous particulates or contaminants. Respirators types vary greatly depending on the intended purpose or environment of use (Table 7.9). Proper usage and fit of respirators is vital to their effectiveness against atmospheric hazards.

Table 7.9 Types of respirator components and properties (Mine Safety and Health Administration (MSHA), 2006a)

Respirator component	Comment/Properties
Facepieces	
Loose-fitting Tight-fitting	Covers the head completely (e.g., hoods or helmets) e.g., half masks (mouth and chin), full facepieces (forehead to chin coverage)
Breathing air	
Air purifying	Filters out contamination. Includes negative pressure, tight-fitting facepieces, cannot be used in oxygen-deficient atmosphere
Supplied air	Provides own source of breathable air. Typically used for unknown or high concentration contaminants, can be positive or negative pressure systems
Pressure	
Positive pressure	Pressure inside respirator is greater than outside, less likely to allow contaminant infiltration
Negative pressure	Pressure inside respirator is less than outside during inhalation

All miners working in an underground mine must have access to a 1-hour self-rescue device for emergency escape. Self-rescue devices are meant to assist in evacuation from contaminated atmosphere (Mine Safety and Health Administration (MSHA), 2006a). These devices can either be worn on the body or on mobile equipment, but must be readily accessible and no more than 25 ft (7.6 m) away from the miner (Mine Safety and Health Administration (MSHA), 2006a). Single-use filter self-rescue devices, typically carried in metal and non-metal mining operations, provide protection against carbon monoxide for up to 60 min, but do not provide a source of oxygen (Mine Safety and Health Administration (MSHA), 2006a). Self-contained self-rescue devices use compressed oxygen or chemical reactions to generate oxygen and can be used in oxygen deficient atmospheres (Mine Safety and Health Administration (MSHA), 2006a).

Refuge Alternatives

The passage of the MINER Act requires the deployment of refuge alternatives at each working face in an underground coal mine. Refuge alternatives provide protection against irrespirable atmospheres, caused by either fires or the release of gases, in underground mining operations. Refuge alternatives primarily consist of either refuge chambers or in-place shelters (National Institute for Occupational Safety and Health (NIOSH), 2007). Refuge alternatives are outfitted to provide food, water, shelter, air, and communications for a set period of time and can vary in size depending on the number of miners working within the area (Table 7.10). It is important to note that refuge alternatives are designed to be a last resort for miners during an emergency, as evacuation usually takes precedence (Kingsley Westerman et al., 2011). Other refuge alternatives, such as mine escape vehicles, are currently under development and have not been fully evaluated (National Institute for Occupational Safety and Health (NIOSH), 2007). Regulations also require refuges in mining operations in other countries, such as Australia (Western Australia Department of Industry and Resources, 2005).

Refuge Chambers

Refuge chambers are usually manufactured rigid or inflatable vessels (National Institute for Occupational Safety and Health (NIOSH), 2007). Refuge chambers are commercially available and are delivered to the mine, where they are moved near the working face (National Institute for

Table 7.10 Recommend parameters and values for refuge alternatives (National Institute for Occupational Safety and Health (NIOSH), 2007)

Parameter	Recommended value or practice
Minimum rated duration	48 h
Strength	15 psi overpressure for 0.2 s
Anchor system	Not recommended at this time
Fire resistance	148 °C (300 °F) for 3 s
Deployment time	Minimize this time when establishing the location of the refuge alternative and consider as part of the travel time
Min. concentration O_2	18.5%
Max. concentration O_2	23%
Max. concentration CO	25 ppm
Gases to be monitored inside chamber	O_2, CO, CO_2
External gases to be monitored	O_2, CO
Max. concentration CO_2	1.0%, not to exceed 2.5% for any 24-h period
Apparent temperature	35 °C (95 °F)
Entry and exit	Provide a means of egress without contaminating the internal environment and/or a means to maintain a safe environment during and after ingress/egress.
Potable water per person	2 to 2.25 qt. per 24 h
Durability	Structurally reinforced and of sufficient physical integrity to withstand routine handling
Purge air volume	No specific recommendation (see entry and exit parameter)
Food, per person	2000 kcal per 24 h
Human waste disposal system	Required
First-aid kit	Required
Occupant-activated annunciation	Battery-powered strobe light or radio homing signal
Communication with surface	Survivable post-disaster system
Minimum distance to working face	305 m (1000 ft)
Maximum distance from working face	Distance that a miner could reasonably travel in 30—60 min, under the expected travel conditions
Security	Visual indication that a refuge alternative has been entered; inspection and maintenance actions required subsequent to discovery

Table 7.10 Recommend parameters and values for refuge alternatives (National Institute for Occupational Safety and Health (NIOSH), 2007)—cont'd

Parameter	Recommended value or practice
Repair materials	Materials and instructions supplied by manufacturer
Testing and approval	Required
Unrestricted floor space	>1.4 m² (15 sq ft) per person
Unrestricted volume	>2.4 m³ (85 cu ft) per person
Capacity	Sufficient to accommodate the maximum number of miners in the area to be served by the refuge alternative

Occupational Safety and Health (NIOSH), 2007). As mining operations expand, chambers must be regularly moved to within the recommended distance from a working face, which, while challenging, was proven to be done both safely and feasibly (Bauer and Kohler, 2009). There is limited data on the number of chamber installed in underground coal mines in the UNITED STATES, but as of 2008, 980 orders were recorder by manufacturers, of which over 90% were of the inflatable type (Bauer and Kohler, 2009).

In-place Shelters

In-place shelters differ from refuge chambers in that they are built on-site and are immobile. In-place shelters are typically constructed by isolating a crosscut using one or more bulkheads, or by mining a cut and installing a bulkhead to isolate the dead-end (National Institute for Occupational Safety and Health (NIOSH), 2007). In-place shelters contain the same supplies and equipment as refuge chambers, but cannot be easily moved to maintain distance with a working face.

Reliable Communication During Emergencies

Emergency communication and tracking (emergency CT) systems necessary to facilitate rescue operations and assess damages after an accident are required in all underground coal mines, as stipulated in the MINER Act. Emergency CT technologies need to be approved by the Mine Safety and Health Administration (MSHA) for safe use in gassy and dust-laden atmospheres that may be present post-accident (Damiano et al., 2014).

A recent survey of 306 underground mines in the United States revealed the dominant use of three technologies for emergency CT systems (Damiano et al., 2014):

- Leaky feeder
- Wireless nodes
- Wired nodes

Regardless of technology, all emergency communication systems must be able to provide all underground personnel with two-way communication with the surface via a wireless medium.

Leaky Feeder Communications

Leaky feeder technology has been commercially available for over two decades and is well known, reliable, and accounts for 44% of emergency communication systems in underground mines in the United States (Damiano et al., 2014). Leaky feeder systems utilize a "leaky" coaxial cable that is specially designed such that radio signals are allowed to radiate out and back in, something a normal coaxial cable is designed to prevent (Laliberté, 2009). Leaky feeder systems are available in both VHF (very high frequency) and UHF (ultra-high frequency) variants (Laliberté, 2009). Leaky feeder systems can utilize commercial radios for two-way communication and can also transmit video and data. In VHF systems, the effective range is relatively short and line of sight is required, causing the need for amplifiers to maintain adequate signal quality (Laliberté, 2009). UHF systems provide greater range and coverage around corners, but may have a higher initial cost (Laliberté, 2009). Because the system relies on coaxial cables for both the transmission and receiving of signals, any damage to the cable or power failure will cause all communications to be lost.

Wired and Wireless Node Communications

Node-based emergency communication systems are the most commonly used systems in underground mines in the United States, accounting for 56% of all mines (Damiano et al., 2014). Node communications systems consist of multiple processor based access points installed in a mine that are all interconnected and can wirelessly link to portable devices for communication (Damiano et al., 2014). The structure of such systems reduces interference and vulnerability to failure, creating a reliable network (Laliberté, 2009). Should any connection between nodes become severed, the signal can reroute through other adjacent nodes. Node communications provide great flexibility in installation, operation, and coverage (Laliberté, 2009).

Of the node systems currently in operation in underground mines in the United States, approximately 60% are wireless node systems (Damiano et al., 2014). Wired systems typically only differ in that connections between nodes use physical wires; however, some wired systems are capable of automatically changing to wireless configurations if physical connections become damaged (Damiano et al., 2014). Node communication systems typically do not rely on wired connections for communication between the underground and surface (Damiano et al., 2014).

Electronic Tracking Systems

Electronic tracking systems are required to determine the location of all underground personnel within a reasonable degree of accuracy (e.g., 200 ft for working areas), and typically require a wireless tracking device for each miner. Tracking systems typically utilize either dedicated radio-frequency identification (RFID) tags, or have tracking abilities built into communication handsets (Damiano et al., 2014).

In RFID systems, each worker has a unique tag, which is then read using a RFID reader to determine the location of the worker (Laliberté, 2009). Passive RFID tag readers power tags wirelessly using a modulated radio-frequency field but have a limited range. Thus, in most mines, RFID tags are battery operated, providing a much greater detection distance (Laliberté, 2009). However, battery operation comes at the risk of power failure.

In cases where the same manufacturer provides both electronic tracking and post-accident communication systems, tracking technology is usually integrated with communication system handsets (Damiano et al., 2014). This method eliminates the need for dedicated tags and readers, instead using signal strength between nodes to determine handset locations (Damiano et al., 2014).

REFERENCES

Argonne National Laboratory, 2008. Technology Demonstration for Reducing Mercury Emissions form Small-Scale Gold Refining Facilities. Retrieved from: http://www.ipd.anl.gov/anlpubs/2008/06/61757.pdf.

Banks, J., 2012. Barriers and Opportunities for Reducing Methane Emissions from Coal Mines. Clean Air Task Force. Retrieved from: http://www.catf.us/resources/whitepapers/files/201209-Barriers_and_Opportunities_in_Coal_Mine_Methane_Abatement.pdf.

Bauer, E., Kohler, J., 2009. Update on refuge alternatives: research, recommendations and underground deployment. Min. Engi. 35, 51.

Bellamy, T., Nicholson, R.T., 2012. Environmental Manual, second ed. District of Columbia Department of Transportation. Retrieved from: http://ddotsites.com/documents/environment/Files/Chapters/Chapter_26_-_Environmental_Mitigation_Basics.pdf.

Blodgett, S., Kuipers, J.R., 2002. Technical Report on Underground Hard-Rock Mining: Subsidence and Hydrologic Environmental Impacts Center for Science in Public Participation. Retrieved from: http://www.csp2.org/files/reports/Subsidence%20and%20Hydrologic%20Environmental%20Impacts.pdf.

Bolan, N., Kunhikrishnan, A., Thangarajan, R., Kumpiene, J., Park, J., Makino, T., Kirkham, M.B., Scheckel, K., 2014. Remediation of heavy metal(loid)s contaminated soils – To mobilize or to immobilize? J. Hazard. Mater. 266, 141–166.

Butler, B.A., 2014. Appendix I – Conventional Water Quality Mitigation Practices for Mine Design, Construction, Operation, and Closure. U.S. EPA, Office of Research and Development. Retrieved from: http://ofmpub.epa.gov/eims/eimscomm.getfile?p_download_id=517009.

Copeland, C., 2014. Mountaintop Mining: Background on Current Controversies Congressional Research Service. Retrieved from: https://fas.org/sgp/crs/misc/RS21421.pdf.

Damiano, N., Homce, G., Jacksha, R., 2014. A Review of Underground Coal Mine Emergency Communications and Tracking System Installations. Coal Age.

Darling, P., 2011. SME Mining Engineering Handbook, third ed. Society for Mining, Metallurgy, and Exploration, Ann Arbor, MI.

Department of Environment Climate Change and Water NSW, 2010. Environmental Compliance and Performance Report: Management of Dust from Coal Mines. Retrieved from: http://www.epa.nsw.gov.au/resources/licensing/10994coalminedust.pdf.

Girard, J.M., 2001. Assessing and Monitoring Open Pit Mine Highwalls, Proceedings of the 32nd Annual Institute of Mining Health. Safety and Research, Salt Lake City, UT.

Gunson, A., Klein, B., Veiga, M., Dunbar, S., 2012. Reducing mine water requirements. J. Clean. Prod. 21, 71–82.

Harris, J., Hunter, D., Birch, P., Short, K., 1987. Vesicular arbuscular mycorrhizal populations in stored topsoil. T. Brit. Mycol. Soc. 89, 600–603.

Harris, M., Weiss, E., Man, C., Harteis, S., Goodman, G., Sapko, M., 2010. Rock dusting considerations in underground coal mines. In: Proceedings of the 13th U.S./North American Mine Ventilation Symposium. MIRARCO-Mining Innovation, Sudbury, Ontario, Canada, pp. 267–271.

He, M., Gong, W., Wang, J., Qi, P., Tao, Z., Du, S., Peng, Y., 2014. Development of a novel energy-absorbing bolt with extraordinarily large elongation and constant resistance. Int. J. Rock Mech. Min. Sci. 67, 29–42.

Hu, X.-m., Cheng, W.-m., Wang, D.-m., 2014. Properties and applications of novel composite foam for blocking air leakage in coal mine. Russ. J. Appl. Chem. 87, 1099–1108.

International Council on Mining & Metals (ICMM), 2012. Water Management in Mining: A Selection of Case Studies. Retrieved from: https://www.icmm.com/document/3660.

Jennings, S.R., Neuman, D.R., Blicker, P.S., 2008. Acid Mine Drainage and Effects on Fish Health and Ecology: A Review. Reclamation Research Group. Retrieved from: http://www.pebblescience.org/pdfs/Final_Lit_Review_AMD.pdf.

Johnson, D.B., Hallberg, K.B., 2005. Acid mine drainage remediation options: a review. Sci. Tot. Environ. 338, 3–14.

Kingsley Westerman, C.Y., McNelis, K.L., Margolis, K.A., 2011. Recommendations for Refuge Chamber Operations Training. National Institute for Occupational Safety and Health (NIOSH), Office of Mine Safety and Health. Retrieved from: http://198.246.124.29/niosh/mining/UserFiles/works/pdfs/2011-178.pdf.

Laliberté, P., 2009. Summary Study of Underground Communications Technologies. CANMET Mining and Mineral Sciences Laboratories. Retrieved from: http://www. wvminesafety.org/PDFs/UndergroundCommunicationsReport.pdf.

Leading Practice Sustainable Development Program for the Mining Industry, 2006. Mine Rehabilitation. Australian Department of Industry, Tourism and Resources. Retrieved from: http://www.industry.gov.au/resource/Documents/LPSDP/LPSDP-MineRehabilitationHandbook.pdf.

Leading Practice Sustainable Development Program for the Mining Industry, 2007a. Biodiversity Management. Australian Department of Industry, Tourism and Science. Retrieved from: http://www.industry.gov.au/resource/Documents/LPSDP/LPSDP-BiodiversityHandbook.pdf.

Leading Practice Sustainable Development Program for the Mining Industry, 2007b. Managing Acid and Metalliferous Drainage. Australian Department of Industry, Tourism and Resources. Retrieved from: http://www.industry.gov.au/resource/Documents/LPSDP/LPSDP-AcidHandbook.pdf.

Leading Practice Sustainable Development Program for the Mining Industry, 2009. Airborne Contaminants, Noise and Vibration. Australian Department of Resources, Energy and Tourism. Retrieved from: http://www.industry.gov.au/resource/Documents/LPSDP/AirborneContaminantsNoiseVibrationHandbook_web.pdf.

Li, Z., Ma, Z., van der Kuijp, T.J., Yuan, Z., Huang, L., 2014. A review of soil heavy metal pollution from mines in China: pollution and health risk assessment. Sci. Tot. Environ. 468–469, 843–853.

Lu, X., Wang, D., Zhu, C., Shen, W., Dong, S., Chen, M., 2015. Experimental investigation of fire extinguishment using expansion foam in the underground goaf. Arab. J. Geosci. 1–9.

Mainiero, R.J., Harris, M.L., Rowland, J.H., 2007. Dangers of toxic fumes from blasting. In: Proceedings of the Annual Conference on Explosives and Blasting Technique. International Society of Explosive Engineers.

Martin Marietta Laboratories, 1987. Dust Control Handbook for Minerals Processing. U.S. Department of the Interior, Bureau of Mines. Retrieved from: https://www.osha.gov/dsg/topics/silicacrystalline/dust/dust_control_handbook.html.

Mine Safety and Health Administration (MSHA), 2006a. Chapter 16: Personal Protective Equipment. U.S. Department of Labor. Retrieved from: http://www.msha.gov/Readroom/HANDBOOK/MNMInspChapters/Chapter16.pdf.

Mine Safety and Health Administration (MSHA), 2006b. Stand Down for Safety: Preventing Fires & Explosions. U.S. Department of Labor. Retrieved from: http://www.msha.gov/S&HINFO/RockDusting/talkptsfireexp.pdf.

Mummey, D.L., Stahl, P.D., Buyer, J.S., 2002. Soil microbiological properties 20 years after surface mine reclamation: spatial analysis of reclaimed and undisturbed sites. Soil Biol. Biochem. 34, 1717–1725.

National Institute for Occupational Safety and Health (NIOSH), 2007. Research Report on Refuge Alternatives for Underground Coal Mines. Retrieved from: http://www.cdc.gov/niosh/docket/archive/pdfs/NIOSH-125/125-ResearchReportonRefugeAlternatives.pdf.

National Institute for Occupational Safety and Health (NIOSH), 2011. A Story of Impact: A Real-Time Monitor to Prevent Coal Dust Explosion Hazards in the Mining Industry. Retrieved from: http://www.cdc.gov/niosh/docs/2011-205/pdfs/2011-205.pdf.

Qin, B.-T., Wang, D.-M., Ren, W.-X., 2005. Study on three-phase foam for preventing spontaneous combustion of coal in goaf. In: Proceedings of the Eight International Symposium on Fire Safety Science. International Association for Fire Safety Science, pp. 731–740.

Queensland Government, 2011. Queensland Guidance Note QGN 20 v. 3: Management of Oxides of Nitrogen in Open Cut Blasting. Retrieved from: https://www.oricaminingservices.com/uploads/Bulk%20Systems/QGN-mgmt-oxides-nitrogen.pdf.

Richard, M., Moher, P., Rossin, R., Telmer, K., 2014. Using Retorts to Reduce Mercury Use, Emissions and Exposures in Artisanal and Small Scale Gold Mining: A Practical Guide. Artisanal Gold Council. Retrieved from: http://www.artisanalgold.org/Retort_guide_Oct2014_lowQ_version1.0_Eng.pdf?attredirects=0&d=1.

Sapko, M.J., Cashdollar, K.L., Green, G.M., 2007. Coal dust particle size survey of US mines. J. Loss Prevent. Proc. 20, 616–620.

Siriwardane, H.J., Kannan, R.S., Ziemkiewicz, P.F., 2003. Use of waste materials for control of acid mine drainage and subsidence. J. Environ. Eng. 129, 910–915.

Spiegel, S.J., Veiga, M.M., 2010. International guidelines on mercury management in small-scale gold mining. J. Clean. Prod. 18, 375–385.

Stahl, P.D., Williams, S.E., Christensen, M., 1988. Efficacy of native vesicular-arbuscular mycorrhizal fungi after severe soil disturbance. New Phytologist 110, 347–354.

Sui, W., Zhang, D., Cui, Z.C., Wu, Z., Zhao, Q., 2014. Environmental implications of mitigating overburden failure and subsidences using paste-like backfill mining: a case study. Int. J. Min. Reclamat. Environ. 1–23.

Telmer, K., Veiga, M.M., 2008. World emissions of mercury from small scale artisanal gold mining and the knowledge gaps about them. In: GMP Presentation, Rome, Italy.

Trevits, M.A., Smith, A.C., Brune, J.F., 2007. Remote mine fire suppression technology. In: Proceedings of the 32nd International Conference of Safety in Mines Research Institutes, pp. 306–312.

U.S. Department of Agriculture, 2000. Heavy Metal Soil Contamination. Retrieved from: http://www.nrcs.usda.gov/Internet/FSE_DOCUMENTS/nrcs142p2_053279.pdf.

U.S. Department of Labor, 2010. MSHA to Publish Emergency Temporary Standard for Rock Dust. Retrieved from: http://www.msha.gov/MEDIA/PRESS/2010/NR100921.pdf.

U.S. Environmental Protection Agency (U.S. EPA), 1995a. Compliation of Air Pollutant Emission Factors: Volume I: Stationary Point and Area Sources – Leadbearing Ore Crushing and Grinding. Retrieved from: http://www.epa.gov/ttn/chief/ap42/ch12/final/c12s18.pdf.

U.S. Environmental Protection Agency (U.S. EPA), 1995b. Compliation of Air Pollutant Emission Factors: Volume I: Stationary Point and Area Sources – Primary Copper Smelting. Retrieved from: http://www.epa.gov/ttn/chief/ap42/ch12/final/c12s03.pdf.

U.S. Environmental Protection Agency (U.S. EPA), 1995c. Compliation of Air Pollutant Emission Factors: Volume I: Stationary Point and Area Sources – Primary Lead Smelting. Retrieved from: http://www.epa.gov/ttn/chief/ap42/ch12/final/c12s06.pdf.

U.S. Environmental Protection Agency (U.S. EPA), 1995d. Compliation of Air Pollutant Emission Factors: Volume I: Stationary Point and Area Sources – Transportation And Marketing of Petroleum Liquids. Retrieved from: http://www.epa.gov/ttn/chief/ap42/ch05/final/c05s02.pdf.

U.S. Environmental Protection Agency (U.S. EPA), 1995e. Compliation of Air Pollutant Emission Factors: Volume I: Stationary Point and Area Sources – Zinc Smelting. Retrieved from: http://www.epa.gov/ttn/chief/ap42/ch12/final/c12s07.pdf.

U.S. Environmental Protection Agency (U.S. EPA), 2000. Technical and Economic Assessment: Mitigation of Methane Emissions from Coal Mine Ventilation Air. Retrieved from: http://www.epa.gov/coalbed/docs/vam.pdf.

U.S. Environmental Protection Agency (U.S. EPA), 2002. EPA Air Pollution Control Cost Manual, sixth ed. Retrieved from: http://www.epa.gov/ttncatc1/dir1/c_allchs.pdf.

U.S. Environmental Protection Agency (U.S. EPA), 2004a. Abandoned Mine Lands Reference Notebook – Appendix C: Current Information on Mine Waste Treatment Technologies. Retrieved from: http://www.epa.gov/aml/tech/appenc.pdf.

U.S. Environmental Protection Agency (U.S. EPA), 2004b. Regulatory Announcement: Clean Air Nonroad Diesel Rule. Retrieved from: http://www.epa.gov/otaq/documents/nonroad-diesel/420f04032.pdf.

U.S. Environmental Protection Agency (U.S. EPA), 2006. Abandoned Mine Lands Case Study – Iron Mountain Mine. Retrieved from: http://www.epa.gov/aml/tech/imm.pdf.

U.S. Environmental Protection Agency (U.S. EPA), 2013. Case Study: Methane Recovery at Non-coal Mines. Retrieved from: http://www.epa.gov/cmop/docs/CMOP-Non coal%20Flyer.pdf.

U.S. Environmental Protection Agency (U.S. EPA), 2014. Case Study: Methane Recovery at Surface Mines in the U.S. Retrieved from: http://www.epa.gov/coalbed/docs/CMOP-Methane-Recovery-Surface-Mines-March-2014.pdf.

U.S. & Fish Wildlife Service, 2012. Conservation Banking: Incentives for Stewardship. Retrieved from: http://www.fws.gov/endangered/esa-library/pdf/conservation_banking.pdf.

United Nations, 2010. Best Practice Guidance for Effective Methane Drainage and Use in Coal Mines. Retrieved from: http://www.unece.org/fileadmin/DAM/energy/se/pdfs/cmm/pub/BestPractGuide_MethDrain_es31.pdf.

Virah-Sawmy, M., Ebeling, J., Taplin, R., 2014. Mining and biodiversity offsets: a transparent and science-based approach to measure "no-net-loss". J. Environ. Manage. 143, 61–70.

Western Australia Department of Industry and Resources, 2005. Guideline: Refuge Chambers in Underground Metalliferous Mines. Safety and Health Division. Retrieved from: http://www.usmra.com/download/MS_GMP_Guide_RefugeChamber.pdf.

Wuana, R.A., Okieimen, F.E., 2011. Heavy metals in contaminated soils: a review of sources, chemistry, risks and best available strategies for remediation. ISRN Ecology 2011.

Wyllie, D.C., Mah, C.W., 2004. Rock Slope Engineering: Civil and Mining, fourth ed. CRC Press.

Yao, Y., Cui, Z., Wu, R., 2012. Development and challenges on mining backfill technology. J. Mater. Sci. Res. 1, 73.

Zhou, F., Ren, W., Wang, D., Song, T., Li, X., Zhang, Y., 2006. Application of three-phase foam to fight an extraordinarily serious coal mine fire. Int. J. Coal Geol. 67, 95–100.

Appendix A: Emission Factors for Air Pollutants Related to Mining and Mineral Processing

Table A1 Default uncontrolled emission factors for air pollutants related to mining processes and mineral processing (Eastern Research Group, 2001)

Process	PM filterable (lbs/unit)	PM$_{10}$ (lbs/unit)	PM condensable (lbs/unit)	So$_x$ (lbs/unit)	NO$_x$ (lbs/unit)	VOC (lbs/unit)	CO (lbs/unit)	Lead (lbs/unit)	Units
Chemical manufacturing									
Sodium carbonate processing									
Solvay process: handling	50	10.5							Tons produced
Monohydrate process: rotary ore calciner: gas-fired	368	24.7		0.01	1.4				Tons produced
Monohydrate process: rotary ore calciner: coal-fired	390	37.1							Tons produced
Rotary soda ash dryers	84	17.6							Tons produced
Fluid-bed soda ash dryers/coolers	146	19							Tons produced
Rotary pre-dryer	3.1	5.2							Tons fed
Bleacher: gas-fired	311	7.8							Tons fed
Rotary dryer: steam tube	67	14							Tons produced
Normal superphosphates									
Grinding/Drying	0.56	4.6							Tons produced
Rock unloading	0.11	0.29							Tons produced
Rock feeder system	0.52	0.06							Tons produced
Mixer/Den		0.27							Tons produced
Curing/Building	7.2	6.1							Tons produced
Triple superphosphates									
Rock unloading	0.14	0.07							Tons produced
Rock feeder system	0.03	0.02							Tons produced
Run of pile: mixer/den/curing	0.03	0.02							Tons produced
Granulator: reactor/dryer	0.1	0.05							Tons produced
Granulator: curing	0.2	0.1							Tons produced

Aluminum ore (bauxite)

Process							Units
Drying oven		0.7	1.4				Tons produced
Fine ore storage		0.0007	3				Tons handled
Aluminum ore (electro-reduction)							
Prebaked reduction cell		54.5	60	0.003	0.1		Tons produced
Horizontal stud Soderberg cell		56.8			1		Tons produced
Vertical stud Soderberg cell					1		Tons produced
Materials handling	10	5.8					Tons produced
Anode baking furnace					1		Tons produced
Roof vents					2.7		Tons produced
Prebake: fugitive emissions		2.9					Tons produced
H.S.S.: fugitive emissions		3.1					Tons produced
V.S.S.: fugitive emissions		3.7					Tons produced
Aluminum hydroxide calcining							
Overall process		24			0.02		Tons produced
By-product coke manufacturing							
Oven charging	0.48		0.02	0.03	2.5	0.6	Tons charged
Oven pushing	1.15	0.5	3.3	0.03	0.2	0.07	Tons charged
Quenching		0.000054		0.6			Tons charged
Coal unloading	0.00011						Tons charged
Oven underfiring	0.47			0.04	2		Tons charged
Oven/Door leaks	0.54	0.51	0.294	0.01	1.5	0.6	Tons charged
Coal crushing	0.11						Tons charged
Coal preheater	3.5	3.4			0.3		Tons charged
Topside leaks		0.08	0.1	0.01	1.5		Tons charged
Combustion stack: coke oven gas	0.47						Tons charged
Combustion stack: blast furnace gas	0.17	0.45	1.08				Tons charged

Continued

Table A1 Default uncontrolled emission factors for air pollutants related to mining processes and mineral processing (Eastern Research Group, 2001)—cont'd

Process	PM filterable (lbs/unit)	PM₁₀ (lbs/unit)	PM condensable (lbs/unit)	Soₓ (lbs/unit)	NOₓ (lbs/unit)	VOC (lbs/unit)	CO (lbs/unit)	Lead (lbs/unit)	Units
Titanium									
Drying titanium sand ore (cyclone exit)	0.5	0.43							Tons processed
Primary copper smelting									
Multiple hearth roaster	45	23.8		280				0.15	Tons processed
Reverberatory smelting furnace after roaster	50	13.6		160				0.072	Tons processed
Converter (all configurations)	36	21.2		740				0.27	Tons processed
Fire (furnace) refining		9.2							Tons processed
Ore concentrate dryer	10	4.8		1					Tons processed
Reverberatory smelting furnace w/ ore charge w/o roasting		13.5							Tons processed
Fluidized bed roaster	55	29.2		360					Tons processed
Electric smelting furnace	100	58		240					Tons processed
Flash smelting	140	83		820					Tons processed
Roasting: fugitive emissions	2.6	1.4		1					Tons processed
Reverberatory furnace: fugitive emissions	0.4	0.17		4					Tons processed
Converter: fugitive emissions	4.4	2.6		130					Tons processed
Anode refining furnace: fugitive emissions	0.5	0.46		0.1					Tons processed
Slag cleaning furnace: fugitive emissions	8	7.7		6					Tons processed
Converter slag return: fugitive emissions				0.1					Tons processed
Slag cleaning furnace	10	7.7		6					Tons processed

Process				
Reverberatory furnace with converter	86	9.7	320	Tons processed
AFT MHR+RF/FBR+EF	36	21.2	600	Tons processed
Fluid bed roaster with reverberatory furnace and converter	86	19.1	360	Tons processed
Dryer with electric furnace and cleaning furnace and convertor	146	17.3	1	Tons processed
Dryer with flash furnace and converter	150	4.8	1	Tons processed
Multiple hearth roaster with reverberatory furnace and converter	131	19	280	Tons processed
Fluid bed roaster with electric furnace and converter	136	19.1	600	Tons processed
Reverberatory furnace after multiple hearth roaster	50	13.5	180	Tons processed
Reverberatory furnace after fluid bed roaster	50	13.5	160	Tons processed
Electric furnace after concentrate dryer	100	58	240	Tons processed
Flash furnace after concentrate dryer	140	83	820	Tons processed
Electric furnace after fluid bed roaster		58		Tons processed
Concentrate dryer followed by Noranda reactors and converter			1	Tons processed
Iron production				
Ore charging	0.026	41.8		Tons produced
Agglomerate charging		15.2		Tons produced
Loader: hi-silt		0.013		Tons transferred

Continued

Table A1 Default uncontrolled emission factors for air pollutants related to mining processes and mineral processing (Eastern Research Group, 2001)—cont'd

Process	PM filterable (lbs/unit)	PM_{10} (lbs/unit)	PM condensable (lbs/unit)	SO_x (lbs/unit)	NO_x (lbs/unit)	VOC (lbs/unit)	CO (lbs/unit)	Lead (lbs/unit)	Units
Loader: low-silt	0.0088	0.0044							Tons transferred
Windbox	11.1	1.67			0.3	1.4	44.7		Tons produced
Discharge end	6.8	1.02							Tons produced
Cooler		0.45		0.14		0.05			Tons produced
Sinter process		0.12							Tons produced
Sinter conveyor: transfer station		0.02							Tons transferred
Unload ore, pellets, limestone, into blast furnace	0.0024	0.0012							Tons transferred
Raw material stockpile: ore, pellets, limestone, coke, sinter						4.8			Tons processed
Blast heating stoves		0.31				0.01			Tons processed
Cast house	0.6			3	0.03	2.8			Tons processed
Blast furnace slips	84	33							Each occurred
Lump ore unloading	0.0003	0.0002							Tons transferred
Unpaved roads: light duty vehicles	1.8	1							Miles travelled
Unpaved roads: medium duty vehicles	7.3	4.1							Miles travelled
Unpaved roads: heavy duty vehicles	14	7.6							Miles travelled
Paved roads: all vehicle types	0.78	0.44							Miles travelled
Lead production									
Sintering: single stream	106.5	208.7		275				105	Tons processed
Blast furnace operation	180.5	321.3		22.5				0.0001	See footnote 1
Dross reverberatory furnace	20	19.6						2.9	Tons processed
Ore crushing	6							0.3	Tons crushed
Materials handling	5	4.25							Tons produced

Sintering: dual stream feed end	213	181	550	174	Tons processed
Slag fume furnace	4.6	1.29	2.9		Tons produced
Lead drossing	0.48	0.47			Tons produced
Raw material crushing and grinding	2.26	0.85			Tons processed
Raw material unloading		0.34			Tons processed
Raw material storage piles		0.26			Tons processed
Raw material transfer		0.43			Tons processed
Sintering charge mixing		1.9			Tons processed
Sinter crushing/screening		0.12			Tons processed
Sinter transfer		0.015			Tons processed
Sinter fines return handling		4.8			Tons processed
Blast furnace tapping (metal and slag)		0.07			Tons produced
Blast furnace lead pouring	0.93	0.93			Tons produced
Blast furnace slag pouring	0.47	0.13			Tons produced
Lead refining/silver retort	1.8	1.76			Tons produced
Lead casting	0.87	0.85			Tons produced
Reverberatory or Kettle softening	3	2.94			Tons produced
Sinter machine leakage	0.68	0.67			Tons processed
Sinter dump area	0.01	0.0008			Tons processed
Sinter machine (weak gas)			550		Tons produced
Taconite iron ore processing					
Primary crushing	0.2				Tons produced
Tertiary crusher	79.8	0.085			Tons produced
Ore transfer	0.1				Tons produced
Bentonite blending	19				Tons stored
Traveling grate feed	0.64				Tons produced
Traveling grate discharge	1.32				Tons produced

Continued

Table A1 Default uncontrolled emission factors for air pollutants related to mining processes and mineral processing (Eastern Research Group, 2001)—cont'd

Process	PM filterable (lbs/unit)	PM₁₀ (lbs/unit)	PM condensable (lbs/unit)	SOₓ (lbs/unit)	NOₓ (lbs/unit)	VOC (lbs/unit)	CO (lbs/unit)	Lead (lbs/unit)	Units
Indurating furnace: gas-fired	29.2	24.8							Tons produced
Indurating furnace: oil-fired	29.2	24.8							Tons produced
Indurating furnace: coal-fired	29.2	24.8							Tons produced
Pellet cooler	0.12								Tons produced
Pellet transfer to storage	3.4	1.5							Tons produced
Haul road: rock	11	6.2							Miles traveled
Haul road: taconite	9.3	5.2							Miles traveled
Bentonite transfer to blending	3.2								Tons transferred
Grate/Kiln furnace discharge	0.82		0.0035						Tons produced
Induration: Grate/Kiln, gas-fired, acid pellets	7.4	0.63	0.022	0.29	1.5		0.014		Tons produced
Induration: Grate/Kiln, gas-fired, flux pellets	7.4	0.63	0.022		1.5		0.1		Tons produced
Induration: Grate/Kiln, gas & oil-fired, acid pellets			0.04						Tons produced
Induration: Grate/Kiln, gas & oil-fired, flux pellets			0.04						Tons produced
Induration: Grate/Kiln, coke-fired, acid pellets				1.9					Tons produced
Induration: Grate/Kiln, coke & coal-fired, acid pellets				2.3					Tons produced
Induration: vertical shaft, gas-fired, acid pellets, top gas stack	16				0.2	0.013	0.077		Tons produced
Induration: vertical shaft, gas-fired, flux pellets, top gas stack	16					0.013			Tons produced

Induration: vertical shaft, gas-fired, acid pellets, bottom gas stack	0.046					Tons produced
Induration: vertical shaft, gas-fired, flux pellets, bottom gas stack	0.046					Tons produced
Straight grate furnace feed					0.63	Tons produced
Straight grate furnace discharge					1.4	Tons produced
Induration: straight grate, gas-fired, acid pellets		0.039	2.5			Tons produced
Induration: straight grate, gas-fired, flux pellets					1.2	Tons produced
Induration: straight grate, oil-fired, acid pellets					1.2	Tons produced
Induration: straight grate, oil-fired, flux pellets						Tons produced
Induration: straight grate, coke-fired, acid pellets		0.039	0.44			Tons produced
Induration: straight grate, coke & gas-fired, acid pellets		0.15				Tons produced
Pellet screen					10	Tons produced
Pellet storage bin loading					3.7	Tons produced
Metal mining (general processes)						
Primary crushing; low moisture ore				0.05	0.5	Tons processed
Secondary crushing; low moisture ore				0.1	1.2	Tons processed
Tertiary crushing; low moisture ore				0.16	2.7	Tons processed
Material handling; low moisture ore				0.06		Tons processed

Continued

Table A1 Default uncontrolled emission factors for air pollutants related to mining processes and mineral processing (Eastern Research Group, 2001)—cont'd

Process	PM filterable (lbs/unit)	PM₁₀ (lbs/unit)	PM condensable (lbs/unit)	So_x (lbs/unit)	NO_x (lbs/unit)	VOC (lbs/unit)	CO (lbs/unit)	Lead (lbs/unit)	Units
Primary crushing; high moisture ore	0.02	0.009							Tons processed
Secondary crushing; high moisture ore	0.05	0.02							Tons processed
Tertiary crushing; high moisture ore	0.06	0.02							Tons processed
Material handling; high moisture ore	0.01	0.004							Tons processed
Dry grinding with air conveying	28.8	26							Tons processed
Dry grinding without air conveying	2.4	0.31							Tons processed
Ore drying	19.7	12			1.6	0.004			Tons processed
Zinc production									
Multiple hearth roaster	227	159							Tons processed
Sinter strand	125	89		0.64					Tons processed
Vertical retort/electrothermal furnace	14.3	93		1.13					Tons processed
Electrolytic processor	6.6	3							Tons processed
Flash roaster	2000	1840		404.4					Tons processed
Fluid bed roaster	2167	1994		223.5					Tons processed
Raw material handling and transfer	3.4	3.4							Tons processed
Sinter breaking and cooling	1.35	1.3							Tons processed
Zinc casting	0.4	2.1							Tons processed
Raw material unloading	0.4	0.23						0.13	Tons processed
Sinter plant, wind box: fugitive emissions	0.24–1.1								Tons produced

								Units
Sinter plant, discharge screens: fugitive emissions	0.56–2.44							Tons produced
Retort building: fugitive emissions	2–4							Tons produced
Casting: fugitive emissions	2.52							Tons produced
Electric retort	20							Tons processed
Leadbearing ore crushing and grinding								
Lead ore w/ 5.1% lead content	6						0.3	Tons processed
Zinc ore w/ 0.2% lead content	6						0.012	Tons processed
Copper ore w/ 0.2% lead content	6.4						0.012	Tons processed
Lead-zinc ore w/ 2% lead content	6						0.12	Tons processed
Copper-lead ore w/ 2% lead content	6.4						0.12	Tons processed
Copper-zinc ore w/ 0.2% lead content	6.4						0.012	Tons processed
Copper-lead-zinc ore w/ 2% lead content	6.4						0.12	Tons processed

Mineral processing

Coal mining, cleaning, and material handling								Units
Fluidized bed	26		0.042	1.4	0.16	0.098		Tons dried
Flash or suspension	16		0.075	0.52				See Footnote 2
Multilouvered	3.7							Tons dried
Unloading	0.02	0.006						Tons shipped
Crushing	0.02	0.006						Tons shipped
Loading		0.05						Tons mined
Hauling		2.1						Miles travelled
Topsoil removal	0.06							Tons removed
Scrapers: travel mode	14.6							Miles travelled

Continued

Table A1 Default uncontrolled emission factors for air pollutants related to mining processes and mineral processing (Eastern Research Group, 2001)—cont'd

Process	PM filterable (lbs/unit)	PM condensable (lbs/unit)	PM$_{10}$ (lbs/unit)	So$_x$ (lbs/unit)	NO$_x$ (lbs/unit)	VOC (lbs/unit)	CO (lbs/unit)	Lead (lbs/unit)	Units
Topsoil unloading	0.04								Tons processed
Overburden	1.3		0.16						Each drilled
Coal seam: drilling	0.22		0.028						Each drilled
Blasting: coal overburden									Cubic yards removed
Dragline: overburden removal	0.06		0.009						
Truck loading; overburden	0.04		0.015						Tons loaded
Truck loading; coal			0.005						Tons loaded
Hauling: haul trucks	17.2		2.1						Miles travelled
Truck unloading: End dump — coal	0.007		0.001						Tons processed
Truck unloading: Bottom dump — coal	0.066		0.01						Tons processed
Truck unloading: bottom dump — overburden	0.002		0.001						Tons processed
Open storage pile: coal			17060						Acre-years existing
Train loading: coal			0.0059						Tons loaded
Bulldozing: coal	49.4								Hours operated
Grading	5.37		3.33						Miles travelled
Overburden replacement	0.012		0.006						Tons processed
Wind erosion: exposed areas	760		380						Acre-years existing
Vehicle traffic: light/medium vehicles	2.79		1.56						Miles travelled

Gypsum manufacturing

Source							Units
Rotary ore dryer	0.16	0.013					Square Feet-Hours flow
Primary grinder/roller mills	2.6	2.2					Tons produced
Not classified	90						Tons throughput
Conveying		0.15					Tons throughput
Primary crushing: gypsum ore	41	0.26					Tons processed
Secondary crushing: gypsum ore	37	1.13					Tons processed
Continuous kettle: calciner	100	26					Tons produced
Flash calciner		14					Tons produced
Impact mill		85					Tons produced
End sawing (8 Ft.)	8	6.8					1000 sq ft sawed
End sawing (12 Ft.)	5	4.25					1000 sq ft sawed

Lime manufacture

Source							Units
Primary crushing	0.017						Tons processed
Secondary crushing/screening	0.62						Tons processed
Calcining; vertical kiln	8	5	8.2	2.8			Tons produced
Calcining; rotary kiln	350	42	6.71	2.8			Tons produced
Calcining; gas-fired calcimatic kiln	97			0.15	0.02	2	Tons produced

Continued

Table A1 Default uncontrolled emission factors for air pollutants related to mining processes and mineral processing (Eastern Research Group, 2001)—cont'd

Process	PM filterable (lbs/unit)	PM$_{10}$ (lbs/unit)	PM condensable (lbs/unit)	So$_x$ (lbs/unit)	NO$_x$ (lbs/unit)	VOC (lbs/unit)	CO (lbs/unit)	Lead (lbs/unit)	Units
Raw material transfer and conveying		0.18							Tons processed
Raw material unloading		0.1							Tons processed
Raw material storage piles		1.4							Tons processed
Product cooler	6.8								Tons produced
Pressure hydrator	0.1	0.07							Tons produced
Product transfer and conveying	2.2								Tons loaded
Calcining: coal-fired rotary kiln	350	42	2.3	5.4	3.1		1.5		Tons manufactured
Calcining: gas-fired rotary kiln					3.5		2.2		Tons manufactured
Calcining: coal- and gas-fired rotary kiln	80	9.6					0.83		Tons manufactured
Calcining: gas-fired parallel flow regenerative kiln				0.0012	0.24		0.45		Tons manufactured
Product loading, enclosed truck	0.61								Tons processed
Product loading, open truck	1.5								Tons processed

Phosphate rock

Process					Units
Drying	5.7	4.8		0.34	Tons processed
Grinding	1.5	0.93			Tons processed
Transfer/Storage	2	1			Tons processed
Open storage	40	14.4			Tons processed
Calcining	15	15			Tons processed
Ball mill	1.46	0.45			Tons milled

Stone quarrying – processing

Process					Units
Open storage		0.12			Ton-years stored
Hauling		6.2			Miles travelled
Drying		5			Tons dried

Potash production

Process					Units
Mine: grinding/drying		13.5			Tons processed

Vermiculite

Process					Units
General			0.47	0.08	Tons processed

Note: Only processes with emission factors are listed in this table. Processes without emission factors were omitted from the list for brevity.

[1]For SO_x, PM_{10}, and PM, filterable, the emission factor units are "Pounds per Tons Concentrated Ore Processed"; for Lead, the emission factor unit is "Pounds per Tons Lead Produced."

[2]For SO_x, the emission factor unit is "Pounds per Tons Wet Coal Dried"; for PM, filterable, the emission factor unit is "Pounds per Tons Coal Dried."

Table A2 Default vehicle exhaust emission factors for various mining equipment (National Pollutant Inventory, 1999)

Equipment	PM₁₀ (kg/unit)	CO (kg/unit)	NOₓ (kg/unit)	SOₓ (kg/unit)	VOCs (exhaust) (kg/unit)	Units
Track type tractor	3.03	9.4	34.16	3.73	3.31	1000 L fuel
Wheeled tractor	5.57	32.19	52.35	3.73	7.74	1000 L fuel
Wheeled dozer	17.7	14.73	34.29	3.74	1.58	1000 L fuel
Scraper	3.27	10.16	30.99	3.74	2.28	1000 L fuel
Grader	2.66	6.55	30.41	3.73	1.53	1000 L fuel
Off-highway truck	17.7	14.73	34.29	3.73	1.58	1000 L fuel
Wheeled loader	3.51	11.79	38.5	3.74	5.17	1000 L fuel
Track type loader	2.88	9.93	30.73	3.74	4.85	1000 L fuel

Table A3 Default uncontrolled emission factors for VOCs from typical degreasing operations (U.S. Environmental Protection Agency (U.S. EPA), 1995b)

Process	Uncontrolled VOC emission factor	Units
Overall	1000	kg/Mg solvent consumed
Cold cleaner		
Entire unit	0.3	Mg/year /unit in operation
Waste solvent loss	0.165	Mg/year /unit in operation
Solvent carryout	0.075	Mg/year /unit in operation
Bath and spray evaporation	0.06	Mg/year /unit in operation
Entire unit	0.4	kg/hr/m^2 surface area and duty cycle
Open top vapor		
Entire unit	9.5	Mg/year/unit in operation
Entire unit	0.7	kg/hr/m^2 surface area and duty cycle
Conveyorized, vapor		
Entire unit	24	Mg/year/unit in operation
Conveyorized, non-boiling		
Entire unit	74	Mg/year/unit in operation

Table A4 Methane emission factors for abandoned gassy coal mines from 1901 according to inventory year and time of closure (Intergovernmental Panel on Climate Change, 2006)

Inventory year	Time interval of mine closure					
	1900–1925	1926–1950	1950–1976	1976–2000	2001–Present
1990	0.281	0.343	0.478	1.561	NA
1991	0.279	0.34	0.469	1.334	NA
1992	0.277	0.336	0.461	1.183	NA
1993	0.275	0.333	0.453	1.072	NA
1994	0.273	0.33	0.446	0.988	NA
1995	0.272	0.327	0.439	0.921	NA
1996	0.27	0.324	0.432	0.865	NA
1997	0.268	0.322	0.425	0.818	NA
1998	0.267	0.319	0.419	0.778	NA
1999	0.265	0.316	0.413	0.743	NA
2000	0.264	0.314	0.408	0.713	NA
2001	0.262	0.311	0.402	0.686	5.735
2002	0.261	0.308	0.397	0.661	2.397
2003	0.259	0.306	0.392	0.639	1.762
2004	0.258	0.304	0.387	0.62	1.454
2005	0.256	0.301	0.382	0.601	1.265
2006	0.255	0.299	0.378	0.585	1.133
2007	0.253	0.297	0.373	0.569	1.035
2008	0.252	0.295	0.369	0.555	0.959
2009	0.251	0.293	0.365	0.542	0.896
2010	0.249	0.29	0.361	0.529	0.845
2011	0.248	0.288	0.357	0.518	0.801
2012	0.247	0.286	0.353	0.507	0.763
2013	0.246	0.284	0.35	0.496	0.73
2014	0.244	0.283	0.346	0.487	0.701
2015	0.243	0.281	0.343	0.478	0.675
2016	0.242	0.279	0.34	0.469	0.652

Table A5 Emission factors for mercury according to various sectors (Spiegel and Veiga, 2010; United Nations Environment Programme and Arctic Monitoring and Assessment Programme, 2008)

Sector	Emission factor (g/unit)	Units
Coal combustion	0.1–0.3	Mg coal
Oil combustion	0.001	Mg oil
Non-ferrous metal		
Copper smelting	5.0	Mg produced
Lead smelting	3.0	Mg produced
Zinc smelting	7.0	Mg produced
Cement production	0.1	Mg produced
Iron and steel production	0.04	Mg produced
Waste incineration		
Municipal waste	1.0	Mg waste
Sewage sludge waste	5.0	Mg waste
Mercury production	200	Mg ore mined
Gold production (large-scale)	0.025–0.027	g gold mined
Gold production (ASGM)		
Whole ore amalgamation	3	g gold mined
Ore concentration amalgamation	1	g gold mined
Ore concentration amalgamation (retort)	0.001	g gold mined
Caustic soda production	2.5	Mg produced

Table A6 Default uncontrolled emission factors for PM$_{2.5}$ emissions related to mining processes and mineral processing (Eastern Research Group, 2001)

Process	PM$_{2.5}$ filterable (lbs/unit)	Units
Primary metal production		
Aluminum ore (electro-reduction)		
Horizontal stud Soderberg cell	39.2	Tons molten aluminum produced
Prebake: fugitive emissions	1.4	Tons molten aluminum produced
H.S.S.: fugitive emissions	1.7	Tons molten aluminum produced
By-product coke manufacturing		
Oven pushing	0.19	Tons coal charged
Coal preheater	2.1	Tons coal charged
Combustion stack: coke oven gas (COG)	0.44	Tons coal charged
Ferroalloy, open furnace		
50% FeSi: Electric smelting furnace	40	Tons material produced
Silicon metal: electric smelting furnace	654	Tons material produced
Silicomanaganese: electric smelting furnace	125	Tons material produced
80% ferromanganese	17	Tons material produced
80% ferrochromium	99	Tons material produced
Iron production		
Windbox	0.56	Tons material produced
Cast house	0.14	Tons material produced
Steel manufacturing		
Open hearth furnace: stack	12.7	Tons material produced
Basic oxygen furnace: open hood-stack	0.0044	Tons material produced
Charging: BOF	0.13	Tons material produced
Tapping: BOF	0.34	Tons material produced

Secondary metal production		
Aluminum		
Smelting furnace/Reverberatory	2.16	Tons metal produced
Fluxing; chlorination	199	Tons metal processed
Mineral products		
Coal mining, cleaning, and material handling		
Fluidized bed	3.8	Tons wet coal dried
Lime manufacturing		
Calcining; coal-fired rotary kiln	4.9	Tons lime manufactured
Calcining; coal- and gas-fired rotary kiln	1.1	Tons lime manufactured

Note: Only processes with emission factors are listed in this table. Processes without emission factors were omitted from the list for brevity.

Table A7 Default emission factors typical mineral processing operations (National Pollutant Inventory, 1999)

Process	PM$_{10}$ for high moisture content ores[1] (kg/unit)	PM$_{10}$ for low moisture content ores (kg/unit)	Units
Primary crushing	0.004	0.02	Metric tons processed
Secondary crushing	0.012	No data	Metric tons processed
Tertiary crushing	0.01	0.08	Metric tons processed
Wet grinding (milling)	0	0	Metric tons processed
Dry grinding with air conveying or classification	13	13	Metric tons processed
Dry grinding without air conveying or classification	0.16	0.16	Metric tons processed
Drying (all minerals except titanium/zirconium sands)	5.9	5.9	Metric tons processed
Handling transferring and conveying (except bauxite)	0.002	0.03	Metric tons processed
Drying titanium/zirconium with cyclones	No data	No data	Metric tons processed

[1]High moisture content is defined at >4% by weight, except for bauxite where high moisture content is 15–18%.

Table A8 Default emission factors and equations to estimate emission factors from various operations at coal mines (National Pollutant Inventory, 1999)

Operation/Activity	TSP equation	PM$_{10}$ equation	TSP emission factor (kg/unit)	PM$_{10}$ emission factor (kg/unit)	Units
Draglines	$E = 0.0046 \times \frac{d^{1.1}}{M^{0.3}}$	$E = 0.0022 \times \frac{d^{0.7}}{M^{0.3}}$	0.06	0.026	m^3
Excavators/Shovels/Front-end loaders (on overburden)	$E = 0.74 \times 0.0016\left(\frac{U}{2.2}\right)^{1.3}\left(\frac{M}{2}\right)^{-1.4}$	$E = 0.35 \times 0.0016\left(\frac{U}{2.2}\right)^{1.3}\left(\frac{M}{2}\right)^{-1.4}$	0.025	0.012	ton
Excavators/Shovels/Front-end loaders (on coal)	$E = 1.56 \times \frac{0.0596}{M^{0.9}}$	$E = 0.75 \times \frac{0.0596}{M^{0.9}}$	0.029	0.014	ton
Bulldozers on coal	$E = 35.6 \times \frac{s^{1.2}}{M^{1.4}}$	$E = 6.33 \times \frac{s^{1.5}}{M^{1.4}}$	101	32	hour
Bulldozer on material other than coal	$E = 2.6 \times \frac{s^{1.2}}{M^{1.3}}$	$E = 0.34 \times \frac{s^{1.5}}{M^{1.4}}$	17	4	hour
Trucks (dumping overburden)	—	—	0.012	0.0043	ton
Trucks (dumping coal)	—	—	0.010	0.0042	ton
Drilling	—	—	0.59	0.31	hole
Blasting	$E = 0.00022 \times A^{1.5}$	$E = 0.000114 \times A^{1.5}$	—	—	blast
Wheel and bucket	—	—			

Continued

Table A8 Default emission factors and equations to estimate emission factors from various operations at coal mines (National Pollutant Inventory, 1999)—cont'd

Operation/Activity	TSP equation	PM$_{10}$ equation	TSP emission factor (kg/unit)	PM$_{10}$ emission factor (kg/unit)	Units
Wheel generated dust	$E = 0.0019 \times A^{1.5}$	—	2	0.4	km traveled
Scrapers	$E = 9.6 \times 10^{-6} \times s^{1.3} \times W^{2.4}$	$E = 1.32 \times 10^{-6} \times s^{1.4} \times W^{2.5}$	2	0.5	km traveled
Graders	$E = 0.0034 \times S^{2.5}$	$E = 0.0034 \times S^{2.0}$	—	—	km traveled
Loading stockpiles	—	—	0.004	0.0017	ton
Unloading from stockpiles	—	—	0.03	0.013	ton
Loading to trains	—	—	0.0004	0.00017	ton
Miscellaneous transfer points	$E = 0.74 \times 0.0016 \times \dfrac{\left(\frac{U}{2.2}\right)^{1.3}}{\left(\frac{M}{2}\right)^{1.4}}$	$E = 0.35 \times 0.0016 \times \dfrac{\left(\frac{U}{2.2}\right)^{1.3}}{\left(\frac{M}{2}\right)^{1.4}}$	0.00032	0.0015	ton
Wind erosion	—	—	0.4	0.2	hectare/hour

Values include combustion generated PM$_{10}$.

D = drop distance; meters

M = moisture content; %

U = mean wind speed; m/s

A = area blasted; m^2

s = silt content; % (by weight)

S = mean vehicle speed; km/h

W = vehicle gross mass; tons

Table A9 Estimated emission control factors for dust control from various operations at mines (National Pollutant Inventory, 1999)

Operation/Activity	Control method	Emission control factor (%)
Coal mining		
Scrapers on topsoil	Operation done when soil is naturally or artificially moist	50
Dozers on coal or other material	No control	—
Drilling	Fabric filters	99
	Water sprays	70
Blasting coal or overburden	No control	—
Loading trucks	No control	—
Hauling	Watering (2 liters/m³/hr)	50
	Watering (>2 liters/m³/hr)	75
Unloading trucks	Water sprays	70
Draglines	Minimize drop heights	—
Loading stockpiles	Water sprays	50
	Variable height stacker	25
	Telescopic chute with water sprays	75
	Total enclosure	99
Unloading from stockpiles	Water sprays	50

Continued

Table A9 Estimated emission control factors for dust control from various operations at mines (National Pollutant Inventory, 1999)—cont'd

Operation/Activity	Control method	Emission control factor (%)
Wind erosion from stockpiles	Water sprays	50
	Wind breaks	30
	Re-vegetation or total enclosure	99
Loading to trains	Enclosure	70
	Enclosure and fabric filters	99
Miscellaneous transfer and conveying	Water sprays with chemicals	90
	Enclosure	70
	Enclosure and fabric filters	99
Metalliferous mines		
General	Windbreaks	30
	Water sprays to keep ore wet	50
	Hooding with cyclones	65
	Hooding with scrubbers	75
	Hooding with fabric filters	83
	Enclosed or underground	100

Table A10 Lead emission factors for ore crushing and grinding (U.S. Environmental Protection Agency (U.S. EPA), 1995a)

Type of ore	Lead emission factor (lb/unit)	Units
Lead	0.002	Ton ore processed
Zinc	0.00008	Ton ore processed
Copper	0.00008	Ton ore processed
Lead–zinc	0.0008	Ton ore processed
Copper–lead	0.0008	Ton ore processed
Copper–zinc	0.00008	Ton ore processed
Copper–lead–zinc	0.0008	Ton ore processed

REFERENCES

Eastern Research Group, 2001. Chapter 14-Uncontrolled Emission Factor Listing for Criteria Air Pollutants, Vol. II. Emission Inventory Improvement Program. Retrieved from: http://www.epa.gov/ttnchie1/eiip/techreport/volume02/ii14_july2001.pdf.

Intergovernmental Panel on Climate Change, 2006. 2006 IPCC Guidelines for National Greenhouse Gas Inventories United Nations. Retrieved from: http://www.ipcc-nggip.iges.or.jp/public/2006gl/pdf/0_Overview/V0_1_Overview.pdf.

National Pollutant Inventory, 1999. Emission Estimation Technique Manual for Mining and Processing of Non-Metallic Minerals. Australian Government Department of the Environment. Retrieved from: http://www2.unitar.org/cwm/publications/cbl/prtr/pdf/cat5/fnonmeta.pdf.

Spiegel, S.J., Veiga, M.M., 2010. International guidelines on mercury management in small-scale gold mining. Journal of Cleaner Production 18, 375–385.

U.S. Environmental Protection Agency (U.S. EPA), 1995a. Compilation of Air Pollutant Emission Factors: Vol. I: Stationary Point and Area Sources – Leadbearing Ore Crushing and Grinding. Retrieved from: http://www.epa.gov/ttn/chief/ap42/ch12/final/c12s18.pdf.

U.S. Environmental Protection Agency (U.S. EPA), 1995b. Compilation of Air Pollutant Emission Factors: Vol. I: Stationary Point and Area Sources – Solvent Degreasing, fifth ed. Retrieved from: http://www.epa.gov/ttnchie1/ap42/ch04/final/c4s06.pdf.

United Nations Environment Programme, Arctic Monitoring and Assessment Programme, 2008. Technical Background Report to the Global Atmospheric Mercury Assessment. Retrieved from: http://www.chem.unep.ch/mercury/Atmospheric_Emissions/Technical_background_report.pdf.

INDEX

Note: Page numbers followed by "f" and "t" indicates figures and tables respectively.

A

Acid mine drainage (AMD), 21–22, 24, 111, 243–244
  chemical and biological reactions, 111–112
  control, 245–247
  environmental and water quality impacts, 115t
  factors affecting, 113
  formation, 112–113
  impacts, 113–115
  mitigation, 243–245
    techniques, 244t
  prediction and modeling, 115–117
  risks of groundwater contamination, 242
  treatment types for control, 246t
Acoustic Doppler Velocimeter, 191
Active devices, 239
Advection-dispersion equation, 106–107
Air crossings, 239
Air doors, 239
Air pollution mitigation and control, 229. *See also* Ecological impact(s)—mitigation; Land impact(s)—mitigation; Water quantity impact(s)—mitigation
  CMM mitigation, 238–239
  mercury emission mitigation, 236–238
  nitrogen oxide emission mitigation, 236
  PM mitigation, 230–234
  sulfur dioxide emission mitigation, 235–236
  VOC emission mitigation, 234–235
Air quality, 53–54, 184–185. *See also* Water quality
  European Union guidelines, 56t
  impacts from mining and mineral processing, 57
    mercury emissions, 75–79
    methane emissions, 58–65

    nitrogen oxides emissions, 73–75
    PM emissions, 79–86
    sulfur dioxide emissions, 66–69
    VOCs emissions, 69–73
  impacts from open and underground mining, 59t
  legislation, 54–57
  modeling impacts, 86–87
    box model, 87–89
    for dust and particulate dispersion, 88t
    Eulerian model, 90–91
    Gaussian model, 89–90
    Lagrangian model, 90
  monitoring, 189
    methane monitoring, 190–191
    PM monitoring, 190
  pollutants affecting, 60t
  primary and secondary standards for criteria pollutants, 55t
  World Health Organization guidelines, 57t
Airlocks, 239
Airtight vessel, 236–237
Amalgam, 75–76
AMD. *See* Acid mine drainage (AMD)
Anthracite coal, 40
Artisanal and small-scale gold mining (ASGM), 75–76, 78–79
ASGM. *See* Artisanal and small-scale gold mining (ASGM)
Audit, 201
Audit methodology, 207
Auditing system checklists, 209
  for EMS audits, 210t–215t
  for mining audits, 216t–226t
Automated particulate emissions control system, 233–234
Automated systems, 193–194

B

BACI. *See* Before-After-Control-Impact method (BACI)
Backward linkages, 138, 143–144
Baghouses, 233
Baseline monitoring design, 164–165, 166t–172t
Before-After-Control-Impact method (BACI), 173
Berlin II Guidelines for Mining and Sustainable Development, 42–44
Best practices (BPs), 35, 39, 45–46
  British Columbia coal mining and aboriginal peoples, 39–41
  costs of implementing, 41–42
Biodiversity net change, 254
Blast furnace, 232
Box model, 87–89
BPs. *See* Best practices (BPs)
British Columbia coal mining and aboriginal peoples, 39–41

C

Canada, 39–40
CEMS. *See* Continuous emissions monitoring system (CEMS)
Chemicals additives, 232
Cinnabar (HgS), 113
Clean Air Act, 54
Clean Water Act (CWA), 97–98
Clear Air Nonroad Diesel Rule (2004), 235–236
CMM. *See* Coal-mine methane (CMM)
Coal dust explosions, 256–258
Coal mines
  gassy, 63–64, 64t
  methane from, 61–65
Coal-mine methane (CMM), 58, 238
  emissions from ventilation air, 238
Community needs, 147–148
Condensers, 235
Conservation banking, 252–254
Consumption linkages, 143
Continuous emissions monitoring system (CEMS), 189
Continuous sampling, 180

Conventional ore processing equipment, 237–238
Covellite (CuS), 113
CWA. *See* Clean Water Act (CWA)

D

Data analysis techniques, 187–188
Data management, 186–187
  data analysis techniques, 187–188
  GIS, 187
Degasification, 58, 238
Detection limits, 179
Deterministic models, 116
Differential interferometric synthetic aperture radar (DInSAR), 196
Direct economic impacts, 138. *See also* Indirect economic impacts
  mining-specific economic impacts, 141–144
  per capita consumption, 138–139
  public sector revenue and expenditures, 139–140
  regional economic stability, 140–141
Direct mining services, 144
Direct suppliers, 143–144
Dissolved oxygen, 193–194
Dry mine seals, 244–245
Dust control technology, 248

E

EA. *See* Environmental assessment (EA)
Ecological conditions, 186
Ecological impact(s), 130. *See also* Air pollution mitigation and control; Land impact(s)—mitigation; Water quality—control
  attributes affected by mining activities, 130t
  mining impacts on, 132–134, 133t
    habitat loss and fragmentation, 134–136
    noise pollution, 136
  mitigation, 251
    mine operation, 254–256
    mine planning, 252–254
    mitigation styles, 253t
  scientific disciplines, 132

Economic impacts, 136–137
  direct impacts, 138–144
  indirect impacts, 144–149
  by mining and mineral processing, 137t
Economic linkages, 142–144
Education impact, 146–147
EIA. *See* Environmental impact
  assessment (EIA)
EIS. *See* Environmental impact statement
  (EIS)
Electronic tracking systems, 265
Electrostatic precipitators, 233
Emergency communication and tracking
  systems, 263–264. *See also* Fire
  and explosion control
  electronic tracking systems, 265
  leaky feeder communications, 264
  wired and wireless node
    communications, 264–265
Emission factors
  lead for ore crushing and
    grinding, 297t
  for mercury, 289t
  methane emission factors for abandoned
    gassy coal mines, 288t
  mineral processing operations, 292t
  to mining processes and mineral
    processing
    uncontrolled for air pollutants, 6t
    uncontrolled for $PM_{2.5}$ emissions,
      290t–291t
    uncontrolled for VOCs, 287t
  vehicle exhausting for mining
    equipment, 11t
  from operations at coal mines,
    293t–294t
  from operations at mines, 295t–296t
Emissions monitoring, 189
Empirical modeling, 116
Employment and wages, 145–146
EMS. *See* Environmental management
  system (EMS)
Engineering controls for dust
  management, 232
Environmental assessment (EA), 17–18,
  22–23
  analysis, 23

Environmental auditing, 201
  audit categories, 202t
  auditing system checklists, 209
  environmental audit subcategories, 203t
  indicators in, 205f
  performance, 202–203
    audit and follow-up completion, 208
    audit initiation, 203–204
    audit plan and activities preparation,
      204–205
    conducting audit, 205–207
    report preparation, 207
  standards for, 208–209
  types, 201–202
Environmental degradation, 35
Environmental impact assessment (EIA),
  17–18
  mining and mineral processing
    projects, 18
  systematic procedure for
    documentation, 18
    action, 19–20
    activities and attributes re-evaluate
      identification, 27
    analyzing findings, 31–33
    developing alternatives, 20–23
    evaluating impacts, 28–30
    examining attributes, 24–26
    functional areas and mining
      activities, 25t
    impact matrix for mining
      project, 29f
    project activities identification, 24
    reviewing alternatives, 31
    soliciting government concerns,
      26–27
    step-by-step procedure, 19f
    summarizing impacts, 30–31
Environmental impact
  statement (EIS), 17
Environmental impacts
  of mining, 53
    air quality impacts, 53–91
    AMD, 111–117
    ecological impacts, 130–136
    economic impacts, 136–149
    land impacts, 117–130

water quality and quantity impacts, 91—111
mitigation, 229
Environmental management system (EMS), 35, 36f, 201
benefits of implementation and BPs, 39
British Columbia coal mining and aboriginal peoples, 39—41
costs of implementing, 41—42
government involvement in, 42—45
implementation, 45—46
conducting reviews and developing improvement plans, 50
EIA and existing process controls, 48
identifying environmental aspects and impacts, 47
identifying environmental requirements, 47
improving controls, 48—49
monitoring and assessing performance, 49—50
policy, 47
project planning, 46—47
ISO 14001, 38
ISO certification, 38—39
managing environmental liabilities and responsibilities, 36
Environmental monitoring
data management, 186—188
design of monitoring plans, 161
air quality, 184—185
baseline monitoring design, 164—165, 166t—172t
ecological conditions, 186
geochemical conditions, 185—186
geotechnical conditions, 185—186
impact monitoring design, 165—184, 174t—177t
monitoring plans, 184
risks associated with, 162—163
water quality, 185
environmental mining performance, 159f
monitoring network creation, 160
process, 159
technologies
air quality monitoring, 189—191
emerging, 196—197
land impact monitoring, 194—196
water quality monitoring, 192—194
water quantity monitoring, 191—192
Environmental policy, 37, 47
Erosion, 125
modeling, 125—126
wind, 81—82
Eulerian model, 90—91
Excel-based deterministic model, 111

F

Fabric filters, 233
Finding of no significant impact (FONSI), 17—18, 32
Fire and explosion control, 256—259. *See also* Emergency communication and tracking systems
foams, 258—259
rock dust, 257—258
Fire prevention, 256—257, 259
Fiscal linkages, 143
Flaring, 235
Foams, 258—259
FONSI. *See* Finding of no significant impact (FONSI)
Forward linkages, 143
Fugitive emissions, 57
Fume hoods, 237
Functional areas, 24, 25t

G

Galena (PbS), 113
Gaseous emissions, 57
Gaussian model, 89—90
Geochemical conditions, 185—186
Geographic information system (GIS), 187
Geotechnical conditions, 185—186
GIS. *See* Geographic information system (GIS)
Global Mercury Project, 237
"Gob" methane, 58, 238
Grab samples, 180
Groundwater, 95, 191—193

H

H$_2$SO$_4$. *See* Sulfuric acid (H$_2$SO$_4$)
Habitat loss and fragmentation, 134—136
Heavy metals, 104—105
    contamination, 118—123
    mitigation, 248—249
    health impacts, 121t—122t
Holistic land management, 252
Human Development Index, 10. *See also*
        Mining Contribution Index (MCI)
Human health impacts, 66—67, 73—74
    due to heavy metal exposure,
        121t—122t
    mining laws, 82
    PM on, 80t
Hydroxyl radicals (OH·), 70

I

Impact(s). *See also* Environmental
        impacts
    evaluation, 28—30
    matrix for mining project, 29f
    monitoring design, 165—173,
        174t—177t
        adaptability and longevity,
            183—184
        indicators, 173
        sampling considerations, 178—183
    summarizing, 30—31
In-place shelters, 263
Income inequality, 146
Indicators, 173
    in environmental audit, 205f
Indirect economic impacts, 144. *See also*
        Direct economic impacts
    community needs, 147—148
    lifestyles, 145—147
    physiological needs, 149
    psychological needs, 148
Indirect producer services, 144
Indirect suppliers, 144
International Organization for
        Standardization (ISO), 37—38,
        208
    ISO 19011, 201, 203
ISO. *See* International Organization for
        Standardization (ISO)

L

Lagrangian model, 90
Land impact(s), 117. *See also* Air
        pollution mitigation and control;
        Ecological impact(s)—mitigation;
        Water quality—control
    erosion, 125—126
    land-use patterns, 128—130
    mitigation
        heavy metal contamination
            mitigation, 248—249
        slope stabilization, 249—250, 250t
        subsidence mitigation, 250—251
        topsoil management, 247—248
    monitoring
        slope stability monitoring,
            195—196
        soil monitoring, 194—195
        subsidence monitoring, 196
    soil contamination, 118—123
    subsidence, 127—128
    topsoil disturbance, 123—125
Land-use patterns, 128—130
Lead particulates, 83—86
Leaky feeder communications, 264
Lifestyles, 145—147
"Like-for-like" habitat, 252—254
Longwall mines, 127

M

Managed species, 131
Mass balance equation, 106
Materials handling plants, 230—231
MCI. *See* Mining Contribution Index
        (MCI)
Mercury
    amalgamation, 76
    emissions, 75
        amalgamation, 76
        ASGM, 75—76, 78—79
        global emissions, 78t
        human exposure, 75
        in mineral processing, 77
        mitigation, 236—238
Methane (CH$_4$), 58
    emission, 58
        from coal mines, 61—65

factors for abandoned gassy coal
        mines, 288t
    mitigation, 238—239
    from non-coal mines, 65
    ventilation, 239
Millerite (NiS), 113
Mine flooding, 244—245
Mine Improvement and New
        Emergency Response Act
        (MINER Act), 259—261, 263—264
Mine operation, 230, 254—256
Mine planning, 230, 252
    conservation banking and offsetting,
        252—254
    mitigation measures, 255t
Mine Safety and Health Administration
        (MSHA), 263—264
MINER Act. *See* Mine Improvement
        and New Emergency Response
        Act (MINER Act)
Mining and mineral processing, 12—13
    challenges, 3—5
    global production for mineral
        commodities, 6t
    hydrometallurgical processing, 13f
    industry trends, 5—8
    ore/mineral processing, 12f
    sustainable development, 1—3
    water quality impacts from, 98—105
Mining Association of Canada, 44—45
Mining Contribution Index (MCI),
        8—10
    mineral export and production
        data and, 11t
Mining subsidence, 127
Mining-specific economic impacts, 141
    economic linkages, 142—144
    transition economies, 141—142
Mitigation measures and control
        technology, 229
    air pollution mitigation and control,
        229—239
    CMM mitigation, 238—239
    mercury emission mitigation,
        236—238
    nitrogen oxide emission
        mitigation, 236

PM mitigation, 230—234
    sulfur dioxide emission mitigation,
        235—236
    VOC emission mitigation,
        234—235
ecological impact mitigation, 251
    mine operation, 254—256
    mine planning, 252—254
    mitigation styles, 253t
fire and explosion control, 256—259
    foams, 258—259
    rock dust, 257—258
human health and safety, 259—260
    emergency communication and
        tracking systems, 263—265
    PPE, 260—261
    refuge alternatives, 261—263
land impact mitigation
    heavy metal contamination
        mitigation, 248—249
    slope stabilization, 249—250, 250t
    subsidence mitigation, 250—251
    topsoil management, 247—248
water quality control
    acid mine drainage, 243—247
    technologies for treating heavy
        metals in mine wastewater, 243t
    water quality impact mitigation,
        241—243
    water quantity impact mitigation, 240
Mitigation procedures, 163
Modeling
    air quality impacts, 86—87
    box model, 87—89
    for dust and particulate
        dispersion, 88t
    Eulerian model, 90—91
    Gaussian model, 89—90
    Lagrangian model, 90
    water quality and quantity, 105—106
        advection-dispersion equation,
            106—107
        excel-based deterministic model, 111
        mass balance equation, 106
Monitoring technologies
    air quality monitoring, 189—191
    emerging, 196—197

land impact monitoring, 194–196
water quality monitoring, 192–194
water quantity monitoring, 191–192
MSHA. *See* Mine Safety and Health
    Administration (MSHA)

N

National Environmental Policy
    Act, 17
National Institute of Occupational Safety
    and Health (NIOSH), 258
National Pollutant Discharge
    Elimination System (NPDES),
    97–98
Net positive impacts, 254
NGOs. *See* Non-governmental
    organizations (NGOs)
NIOSH. *See* National Institute of
    Occupational Safety and Health
    (NIOSH)
Nitric oxide (NO), 73
Nitrogen dioxide (NO_2), 73
Nitrogen oxides (NO_x), 70
    emissions, 73–75
    mitigation, 236
NO. *See* Nitric oxide (NO)
"No-action" alternative, 23
Node communications systems, 264
Node-based emergency
    communication systems, 264
Noise pollution, 136
Non-coal mines, methane emissions
    from, 65
Non-governmental organizations
    (NGOs), 39
Non-mining industries, 44
Nondestructive sampling and
    tracking, 197
NPDES. *See* National Pollutant
    Discharge Elimination
    System (NPDES)

O

Offsetting, 252–254
Original equipment manufacturers
    (OEMs), 143
Ozone (O_3), 70

P

Parametric monitoring system, 189
Particle emissions, 57
Particulate matter (PM), 53–54, 79
    chemical constituents, 80t
    collection systems, 233
    collector systems, 234t
    control technology, 232–234
    emissions, 79, 81–82
        health effects, 80–81
        on human health and
            environment, 80t
        lead particulates, 83–86
        mining processes, 82
        nonspecific emission
            factors, 82–83
    mitigation of, 230–234
    monitoring, 190
    values for areas in near surface
        coal mines, 83t
Passive devices, 239
Passive sampling, 180
Passive treatments, 246–247
Per capita consumption,
    138–139
Personal protective
    equipment (PPE),
    260–261
Physiological needs, 149
Phytoextraction, 249
"Plan-do-check-act" model of
    management, 36
PM. *See* Particulate matter (PM)
Policy, 47
Polyvinyl chloride (PVC), 192
Post-drain methane, 58
PPE. *See* Personal protective
    equipment (PPE)
Practical alternative, 23
Pre-drain methane, 58
Project planning, 46–47
Psychological needs, 148
Public scoping process, 26
Public sector revenue and
    expenditures, 139–140
PVC. *See* Polyvinyl chloride (PVC)
Pyrite (FeS_2), 112

Q

Quality assurance and quality control program (QA/QC program), 161, 181–183

R

Radio-frequency identification (RFID), 265
Real-time monitoring, 178–179
"Reduce, reuse, recycle", 240, 241t
Refuge alternatives, 261
  in-place shelters, 263
  parameters and values, 262t–263t
  refuge chambers, 261–263
Regional economic stability, 140–141
Regulators, 239
Remote sensing, 181
  methods, 197
Remote surveillance, 181
Respirators, 260
Retort, 236–237
RFID. See Radio-frequency identification (RFID)
Risk management framework, 163
Risk register. See Risk management framework
Roasting, 231
Rock dust, 257–258
Room-and-pillar mines, 127

S

Sampling considerations, 178
  detection limits, 179
  QA/QC program, 181–183
  real-time monitoring, 178–179
  sampling sites and methods selection, 179–181
Sampling site selection, 179–181
Seals, 239
Semi-continuous sampling, 180
Sensor calibration, 182
Shotcrete, 249–250
Sintering, 231–232
Slope stability monitoring, 195–196
Slope stabilization, 249–250, 250t
Soil contamination, 118
  heavy metal contamination, 118–123

Soil monitoring, 194–195
Species diversity, 131
Standards for environmental audit, 208–209
Stockpiles, 231
Stockyards, 231
Stoppings, 239
Stream Buffer Zone Rule, 252
Subsidence, 127
  environmental effects, 128
  mitigation, 250–251
  monitoring, 196
  room-and-pillar mines, 127
Sulfur dioxide (SO_2), 66
  emissions, 66–69
    from copper, lead, and zinc smelting, 69t
    in United States, 68f
  health and environmental effects, 67t
Sulfur dioxide emission mitigation, 235–236
Sulfuric acid (H_2SO_4), 235
  plants, 235
Suppression systems, 232
Suspended solids, 100–103
  on surface water, 103t
System stability, 131

T

Tailings storage facilities (TSF), 240
Teragrams of carbon (TgC), 71
Topsoil disturbance, 123–125
Topsoil management, 247–248
Transfer plants, 230–231
Transition economies, 141–142
TSF. See Tailings storage facilities (TSF)

U

Ultra-high frequency variant (UHF variant), 264
Underground storage tank filling methods, 234–235
Universal Soil Loss Equation, 125

V

Vehicular NO_x emission mitigation, 236
Vehicular SO_x emission mitigation, 235–236

Very high frequency variant (VHF variant), 264
Volatile organic compounds (VOCs), 56—57
  emissions, 69—73, 235
    from cleaning and degreasing operations, 73
    global emissions of non-methane, 72f
    mitigation, 234—235
    sources, 72

W
Water, 91—92
  acid mine drainage environmental and water quality impacts, 115t
  environmental impacts on quality attributes, 101t—102t
  factors affecting AMD potential and generation rates, 114t
  heavy metals, 104—105
  and mass balance modeling, 109t—110t
  mine water types, 94t
  modeling quality and quantity, 105—106
  primary and secondary maximum contaminant levels, 99t
  quality
    impacts from mining and mineral processing, 98—105
    legislation, 97—98
  quantity monitoring, 191—192
  suspended solids, 100—103
  uses and management considerations at mining, 93t
  water based foams, 259

Water Erosion Prediction Project (WEPP), 126
Water quality, 185. *See also* Air quality; Air pollution mitigation and control; Ecological impact(s)—mitigation; Land impact(s)—mitigation
  control
    acid mine drainage, 243—247
    technologies for treating heavy metals in mine wastewater, 243t
    water quality impact mitigation, 241—243
  legislation, 97—98
  monitoring, 192—194
Water quantity impacts, 92
  groundwater, 95
  mine water types, 94t
  mining operations, 96—97
  mitigation, 240
  uses and management considerations at mining, 93t
WEPP. *See* Water Erosion Prediction Project (WEPP)
WEQ. *See* Wind Erosion Equation (WEQ)
Wet scrubbers, 233
Wildlife management, 131
Wind Erosion Equation (WEQ), 126
Wired and wireless node communications, 264—265

Y
"Yellow boy", 115

Printed in the United States
By Bookmasters